Butterworths Te
Scientific Checkb

Geotechnics 4 Checkbook

Roy Whitlow

Butterworths
London Boston Durban Singapore Sydney Toronto Wellington

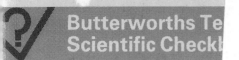

First published 1983

British Library Cataloguing in Publication Data

Whitlow, R.
Geotechnics 4 checkbook.
1. Soil mechanics
I. Title
624.1'5136 TA710

ISBN 0-408-00676-5
ISBN 0-408-00631-5 Pbk

Typeset by Scribe Design Ltd, Gillingham, Kent
Printed in Scotland by Thomson Litho Ltd, East Kilbride
Bound by Hunter & Foulis Ltd, Edinburgh

Contents

Note to Reader

As textbooks become more expensive, authors are often asked to reduce the number of worked and unworked problems, examples and case studies. This may reduce costs, but it can be at the expense of practical work which gives point to the theory.

Checkbooks if anything lean the other way. They let problem-solving establish and exemplify the theory contained in technician syllabuses. The Checkbook reader can gain *real* understanding through seeing problems solved and through solving problems himself.

Checkbooks do not supplant fuller textbooks, but rather supplement them with an alternative emphasis and an ample provision of worked and unworked problems. The brief outline of essential data–definitions, formulae, laws, regulations, codes of practice, standards, conventions, procedures, etc–will be a useful introduction to a course and a valuable aid to revision. Short-answer and multi-choice problems are a valuable feature of many Checkbooks, together with conventional problems and answers.

Checkbook authors are carefully selected. Most are experienced and successful technical writers; all are experts in their own subjects; but a more important qualification still is their ability to demonstrate and teach the solution of problems in their particular branch of technology, mathematics or science.

Authors, General Editors and Publishers are partners in this major low-priced series whose essence is captured by the Checkbook symbol of a question or problem 'checked' by a tick for correct solution.

Preface

The aim of this book is to provide a concise text for use by students studying the TEC unit Geotechnics 4 as part of Civil Engineering or Building courses.

The essential information and principles in appropriate topic areas of geology and soil mechanics are set out in a simple and straightforward manner. Worked examples are included in the text to illustrate the application of concepts and theories to practical problem solving. At the end of each section a generous number of test exercises is given for completion by the student during the progression of the course. These should provide a means of assessing progress both for the student and the teacher, as well as functioning as learning reinforcement. Answers are given at the end of the book.

The author wishes to stress that this text is intended to be used in conjunction with a normal type of lecturer/tutorial/laboratory learning scheme. It is not meant to replace organised teaching/learning experiences, nor the function of a teacher. It is intended to assist the teacher and to provide the student with basic information and a means of self-assessment.

Roy Whitlow
Bristol Polytechnic

Acknowledgement
Extracts from British Standards are reproduced by permission of the British Standard Institution, 2 Park Street, London, W1A 2BS, from whom complete copies and a list of other publications may be obtained.

Butterworths Technical and Scientific Checkbooks

General Editor for Building, Civil Engineering, Surveying and Architectural titles:
Colin R. Bassett, lately of Guildford County College of Technology.

General Editors for Science, Engineering and Mathematics titles:
J.O. Bird and A.J.C. May, Highbury College of Technology, Portsmouth.

A comprehensive range of Checkbooks will be available to cover the major syllabus areas of the TEC, SCOTEC and similar examining authorities. A comprehensive list is given below and classified according to levels.

Level 1 (Red covers)
Mathematics
Physical Science
Physics
Construction Drawing
Construction Technology
Microelectronic Systems
Engineering Drawing
Workshop Processes & Materials

Level 2 (Blue covers)
Mathematics
Chemistry
Physics
Building Science and Materials
Construction Technology
Electrical & Electronic Applications
Electrical & Electronic Principles
Electronics
Microelectronic Systems
Engineering Drawing
Engineering Science
Manufacturing Technology
Digital Techniques
Motor Vehicle Science

Level 3 (Yellow covers)
Mathematics
Chemistry
Building Measurement
Construction Technology
Environmental Science
Electrical Principles
Electronics
Microelectronic Systems
Electrical Science
Mechanical Science
Engineering Mathematics & Science
Engineering Science
Engineering Design
Manufacturing Technology
Motor Vehicle Science
Light Current Applications

Level 4 (Green covers)
Mathematics
Building Law
Building Services & Equipment
Construction Technology
Construction Site Studies
Concrete Technology
Economics for the Construction Industry
Geotechnics
Engineering Instrumentation & Control

Level 5
Building Services & Equipment
Construction Technology
Manufacturing Technology

1 Geological history and time

1.1 EARTH: ORIGIN, AGE AND STRUCTURE

The precise origin and age of planet Earth will probably always remain a mystery. From the time of the early Greek philosophers man has pondered on this question, and although present-day estimates are more reliable, nevertheless they are still estimates and subject to much controversy and contentious argument. It is generally accepted at the present that the age of the Earth is approximately 4600 million years (MY) – give or take a hundred million years.

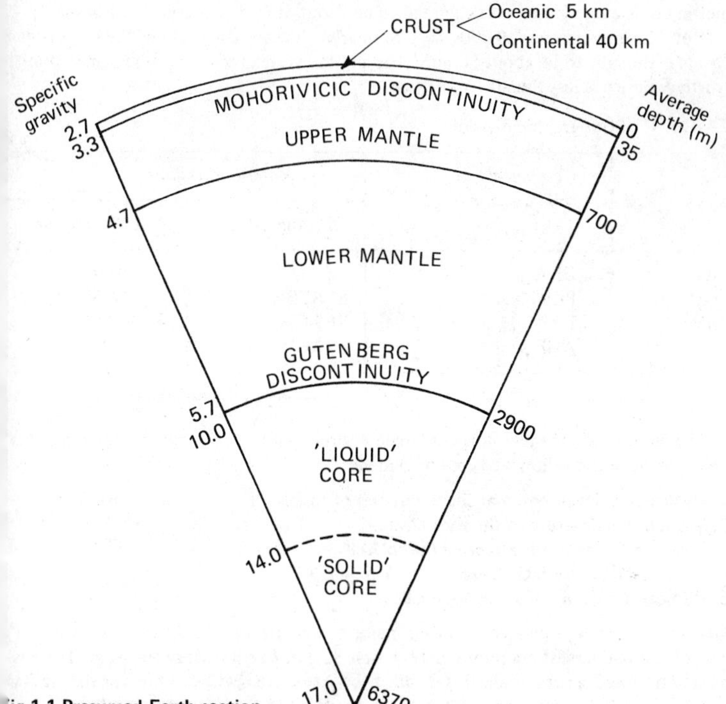

Fig 1.1 Presumed Earth section

The first 1000 MY is thought to have been an accretion stage, during which materia not captured by our sun was gathered up and condensed into a coherent spheroidal body. As compression took place and as radioactive disintegration developed much heat would have been generated; thus the material began to melt. Gravitational forces, together with this melting, then caused the heavier nickel and iron material to fall to the centre forming a dense **core**, leaving lighter material to form the **mantle** (*Fig 1.1*). The first rocks were formed as the outer skin began to cool and distort; the oldest rocks that have been found today are approximately 3700 MY old.

So the Earth (*Fig 1.1*) consists of a dense core, which is thought to be solid at the very centre, the bulk of material forming the mantle, and with a relatively thin outer layer of very light and re-worked material termed the **crust**. From the observation of seismic waves generated by earthquakes, two very definite material divisions have beer located:

Mohorovicic discontinuity – between crust and mantle.
Gutenberg discontinuity – between mantle and core.

1.2 THE GEOLOGICAL TIME SCALE

Geological processes are in the main very slow, but then the periods of time involved are very large. For example, the uplift that has occurred in Scandinavia following the melting of Ice Age glaciers is estimated to be about 500 m, achieved at an average rate of 12 mm/year over 40 000 years. Similarly, the spreading rate of the great ocean basins is thought to be about 50 mm/year and the vertical erosion rate estimated in North America is less than 0.1 mm/year.

TABLE 1.1 Stratigraphic division

Rank	*Geological time division*	*Material divisions*	
		Stratigraphic	*Lithographic*
1st	ERA		GROUP
2nd	PERIOD	SYSTEM	FORMATION
3rd	EPOCH	SERIES	MEMBER
4th	AGE	STAGE	BED

(Stratigraphic and Lithographic: not equivalent)

The geological time record is contained entirely within the rocks themselves and is deduced from the following types of evidence:

1 Mature and developmental characteristics of fossils.
2 Sequence and nature of the rock strata.
3 Structural formations affecting the rocks.
4 Chemical and physical changes in rock minerals.
5 Radioactive decay of some minerals.

Geological history is divided into four major bands of time called **Eras** (see *Table 1.1*) The oldest and largest era represents the first 80% of Earth history for which there is almost no fossil record at all; this is called the **Azoic** era ('without life') or the **Archae** era ('ancient'). The remaining 600 MY are divided into three (unequal) eras as follows

alaeozoic ('ancient life') or Primary ('first')
lesozoic ('middle life') or Secondary ('second')
'ainozoic ('new life') or Tertiary ('third')
Quaternary ('fourth')

:ras are thus major time divisions, which themselves are divided into narrower bands of ime called **Periods,** these in turn being divided into still narrower bands called **Epochs** nd further into **Ages** (*Table 1.1*). Since the rocks laid down during a particular band f time represent the present-day record of that time, it is appropriate to give rock trata **time-based** names which locate them in a chronological sequence.

The time-order of the strata, or the order in which they were laid down, is called the **tratigraphic succession**. Groups of strata are given **stratigraphic** names according to heir age. In *Table 1.1,* these are related in ranking order to the time bands themselves. o the strata laid down during a **Period** is called a **System,** the strata laid down in an **:poch** is called a **Series** and that in an **Age** a **Stage**. **Stratigraphic** names therefore efine the age of strata and are conventionally written as proper nouns with capital etters, e.g. **Devonian** system, **Carboniferous Limestone** series. The rock material in a articular stratum is described using a material or **lithographic** name, written as an rdinary noun (i.e. no capital letter), e.g. limestone, shale, sandstone.

'Telling the time' in geological terms means knowing the chronological order of the ystems or periods shown in *Table 1.2.* It will therefore be necessary for the student to earn these off by heart.

.3 STRATIGRAPHY

tratigraphy is the business of arranging strata or geological events in chronological rder. The boundaries on a geological map are commonly chosen to represent divisions etween stratigraphic series, and the names given are stratigraphic names. The actual ock material may change across a given area beacuse of changes in the conditions revailing at the time of deposition, the stratigraphy in this case would probably be ased on the fossil content of the rocks, which is generally more consistent with time.

The sequence of strata of the same age may be correlated between different areas nce the correct stratigraphic identification has been established. For example, rocks f the Carboniferous Limestone series are of the same age wherever they may be found nd whatever the actual rock material (limestone, shale, or sandstone, etc.) may be.

The rule for placing strata in their correct stratigraphic sequence is called the **'rinciple of Superposition**. This states simply that an upper layer is younger than a ower layer, unless they have been inverted due to crustal movement. Another way of utting it is to say that the top of a bed is younger than the bottom. In steeply inclined nd overturned strata, 'top' and 'bottom' may be defined using **way-up criteria**, e.g.

urface marks	– raindrop prints, tracks, footprints, trails, flow marks, ripple marks.
urface borings	– by bottom-living molluscs, etc.
:ross-bedding	– or current-bedding, see *Fig 1.2.*
'ossil growth positions	– growth of coral, plants, roots, etc.
:hanges in material	– e.g. if the water was getting shallower coarser material would occur nearer the top, or the upper surface of a lava flow may have cooled more quickly and therefore have a finer texture.

TABLE 1.2 Geological Time Scale

Era	Period	Epoch	Approx. Age (MY) Beginning (Duration)	Principal formations and series in UK	Major events
Cainozoic	Quaternary	Recent	0.01	Alluvium, beach deposits, peat, soil	Early civilisation
		Pleistocene	2	Glacial drift (boulder clay, moraine sands & gravels) East Anglian Crags (shelly sands & clays)	Great ice age *Homo Sapiens*
	Tertiary	Pliocene	7 (5)	Coralline Crag (E. Anglia) (Shelly sands) Lenham Beds (Kent, Sussex) (sands)	Primates
		Miocene	26 (19)	None in U.K.	ALPINE OROGENY
		Oligocene	38 (12)	Headon & Hamstead Beds (Hants & I.O.W.) (marls, sandy clays, shelly 1st)	Alps, Pyrenees, Carpathians, Himalayas
		Eocene	54 (16)	Barton & Bracklesham Beds (sands, clays) Bagshot Sands. London Clay	Wealden folding: I.O.W. London Basin
		Palaeocene	65 (11)	Woolwich and Reading Beds (sands, clays) Thanet sands.	Volcanoes in N.W.Scotland Mammals, shellfish 'modern' plants
Mesozoic	Cretaceous		136 (71)	Upper: Chalk (marine lsts) Upper Greensand (ssts, marls, siltsts) Lower: Lower Greensand (marine sands & clays) Weald Clay Hastings Beds (sands, silts, clays)	General uplift Flowering plants Early mammals and birds
	Jurassic		190 (54)	Upper: Purbeck Beds (limestones, marls) Portland Beds (limestones) Kimmeridge Clay Corallian Beds (1sts & calcareous ssts) Oxford Clay. Kellaways Beds (clays, sands)	Conifers, ferns Reptiles dominant
				Middle: Great Oolite (oolitic 1st) Fullers Earth Clay. Inferior Oolite	
				Lower: Upper Lias (shales, sands) Middle Lias (calc. sst & ironstone) Lower Lias (clays, 1sts, shales, ironstone)	Ammonites brachiopods & echinoderms in

	Triassic	225 (35)	Rhaetic (shales, thin 1sts) Keuper marl (red siltstone with salts) Keuper sandstone Bunter Sandstone. Bunter Pebble Beds Dolomitic Conglomerate (scree breccia)	Development of reptiles Marine transgression
Upper Palaeozoic	Permian	280 (55)	Upper Permian Marl Magnesian Limestones (1st, dolomite with salts) Lower Permian Marl and Sands	Sea level falling HERCYNIAN OROGENY Pennines, Mendips, Malverns
	Carboniferous	345 (65)	Coal Measures (shales & ssts with coal seams) Millstone Grit (gritstones & conglomerates) Carboniferous Limestone (1sts, shales, ssts)	Cornish Granite. N. Eng. sills & dykes Mid. Scot. lavas
	Devonian	395 (50)	Lower, Middle & Upper Old Red Sandstones (ssts, conglomerates, marls) Marine Devonian (ssts, shales, 1sts): e.g. Pilton Beds, Baggy Beds, Morte Slates, Ilfracombe Beds, Torquay Lst, Dartmouth Beds.	Fishes, insects Seed plants
Lower Palaeozoic	Silurian	440 (45)	Ludlow Series (mudsts, siltsts, grits) Wenlock Series (shales, 1sts, grits greywackes) Llandovery Series (mudsts, shales, ssts)	CALEDONIAN OROGENY Lake Dist., Wales, S. Scot, Moine Thrust. Granite in Aberdeen, Cairngorm, Cheviot, Skiddaw.
	Ordovician	500 (60)	Ashgill Series (shales, ssts) Caradoc Series (volcanics, shales, 1sts) Llandeilo Series (shales, 1sts, flags) Llanvirn Series (shales, lavas, tuffs) Arenig Series (shales, slates, lavas, tuffs)	Lavas & tuffs in S. Scot. & Mendips. Trilobites, brachiopods, corals, molluscs
	Cambrian	570 (70)	Durness Limestone Tremadoc Series (mudsts, shales, slates) Dolgelly & Ffestiniog Beds (ssts, siltsts) Islay Quartzite & Limestone Llanberis Slates; Harlech Grits	Planktonic algae Graptolites & other invertebrates appear. Uplift in Britain Sea floor sinking
Eozoic	Pre-Cambrian		Torridonian (ssts, arkoses) Lewisian (schists, igneous gneisses, b & u/b dykes) Moinian (schists, granulites) Dalradian (schists, granulites) Uriconian (lavas & tuffs in Salop) Cornish (schists); Mona (schists, quartzites)	CHARNIAN OROGENY Folding in Salop, Scot. Metamorphism in Scot. Volcanics in Wales. Stromatolites & blue-green algae.

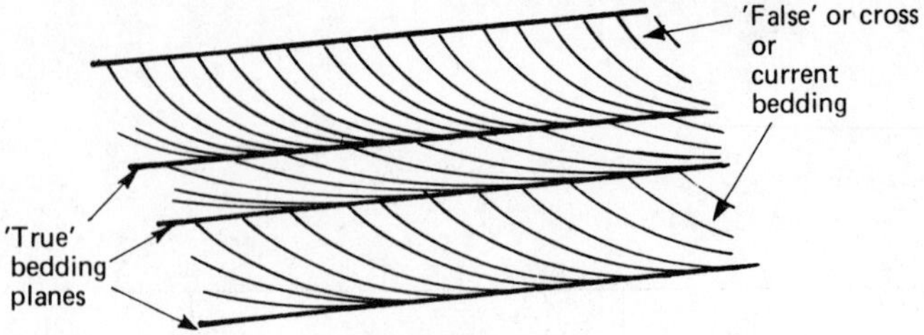

Fig 1.2 Cross-bedding

1.4 STRATIGRAPHY OF THE BRITISH ISLES

Rocks may be found in the British Isles to represent all of the geological time scale periods, with the exception of the Miocene. The oldest rocks are of Pre-Cambrian age and occur in NW Scotland and in Angelsey.

Rocks of the Tertiary period are found in East Anglia and the London and Hampshire basins, and various materials of the Pleistocene occur in areas which underwent glaciation, i.e. most of the country, except southern and south western counties. The student is advised to study the maps published by the Institute of Geological Sciences, see Chapter 4.

TEST EXERCISES

COMPLETION QUESTIONS (answers on page 204)

Complete the following statements by inserting appropriate words in the spaces indicated.

1 The age of the Earth is approximately . . . . million years.

2 The Earth consists of three principal parts: a central . . . . surrounded by the greatest portion called the . . . . , with an outer skin called the . . . .

3 The Gutenberg discontinuity separates the . . . . from the . . . . , and the Mohorivicic discontinuity separates the . . . . from the . . . .

4 Geological time is divided into five major divisions called . . . . and successively smaller divisions called . . . . . . , ... and . . . .

5 State the geological periods in their correct chronological order.

6 The order in which rock strata are laid down is referred to as the . . . . succession.

7 A set of strata laid down during a Period of time is known as a . . . . and that laid down during an Epoch a . . . .

8 Identify the time periods to which the following formations belong: Keuper (. . . . . (. . . . .), Coal Measures (. . . . .), Chalk (. . . . .).

9 The rule for non-inverted sequences of strata which states that the uppermost members are the youngest is called the . . . . of . . . .

10 State *three* 'way-up' criteria which may be used to distinguish the top of a bed from the bottom: . . . . , ... , ...

11 The oldest rocks in the British Isles occur in . . . . and . . . .

12 The youngest rocks, i.e. those of the . . . . Period, occur in . . . .

ESSAY QUESTIONS

1 Draw a diagrammatic cross-section of the Earth and mark on it the main zones and boundaries, together with the supposed densities and depths.

2 Discuss the types of evidence available from which the main divisions of geological time are deduced.

3 Draw up a geological time scale for the strata occurring in the British Isles and locate on this the ages of the strata in the vicinity of your own home.

4 Discuss the geomorphological and economic importance of the Carboniferous Period in Great Britain.

5 Explain what is meant by the term **stratigraphic succession** and describe methods of indicating this on maps and in reports.

2 Surface processes

2.1 CONTINUAL CHANGE AND THE GEOLOGICAL CYCLE

The Earth may be thought of as a continuously running and continually changing dynamic system. It is a restless machine made up of a complex mixture of processes that are forever re-working the materials at the crustal surface. It is also a self-perpetuating system: as material is removed from land masses due to weathering and erosion, it is deposited elsewhere, where it may be formed into new rocks; these in turn may become future land masses and thence be eroded and so on.

The surface of the Earth is constantly shifting and changing: the continents drift slowly, new rocks are formed by volcanic action, mountains crumble, coastlines recede and new sediments are deposited. Wind, rivers, ice and the sea wear away at the land and repeatedly remould and reshape the landscape. It is an everchanging scene.

The cyclic nature of geological processes was first recognised in 1785 by James Hutton, a Scottish farmer and medical practitioner interested in geology. *Fig 2.1* shows

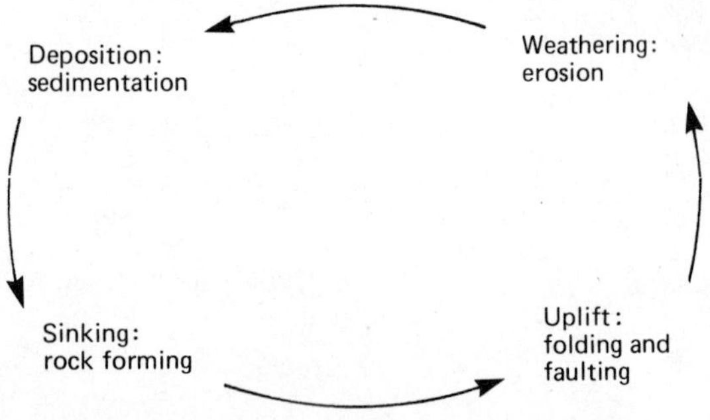

Fig 2.1 Simple geological cycle (After Hutton)

Hutton's basic geological cycle. *Fig 2.2* shows a more comprehensive cycle involving other geological processes and their relationships with the materials of the Earth's crust. The arrows indicate some of the possible paths of change that could lead to the formation of certain materials. The diagram is not meant to be definitive, nor is it exhaustive, it is merely intended as a guide and an illustration of the cyclic nature of earth processes.

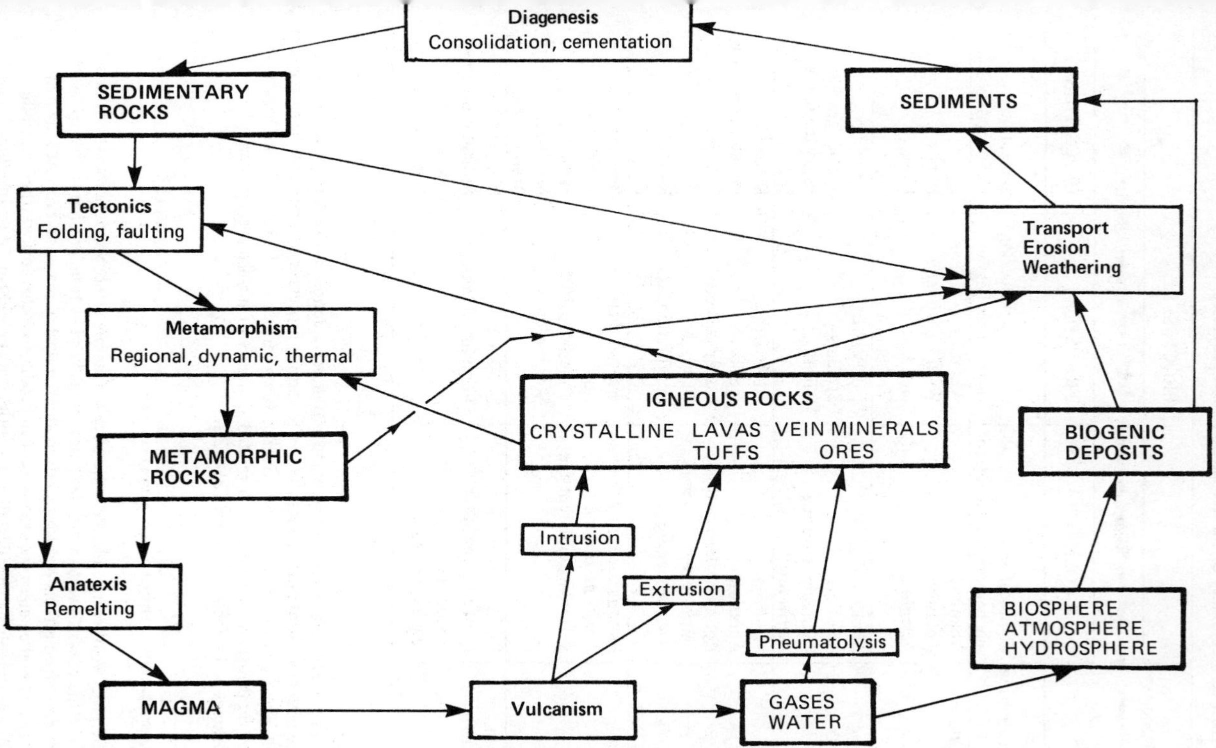

Fig 2.2 Cycle of geological processes and products

TABLE 2.1 Summary of surface agencies and processes

AGENCY	PROCESSES		
	Wearing down	*Building up*	
	EROSION	TRANSPORT	DEPOSITION
Weathering	Chemical and mechanical disintegration	Gravity, solifluction (and others below)	Scree, talus Residual deposits
Wind	Ablation Abrasion Attrition	AEOLIAN Wind-blown sand and dust Volcanic ash	Dunes Loess Tuff
Rivers	Abrasion, cutting Solution Hydraulic scouring Attrition	FLUVIAL Streams Rivers	Alluvium Terrace gravels Estuarine sands Deltaic sands
Ice	Abrasion Rounding & smoothing Plucking & scratching Over-deepening	GLACIAL Ice-sheets Valley glaciers Outwash Solifluction	DRIFT Boulder clay (till) Drumlins Moraines, Eskers Kames Head
Sea	Bursting, scouring Abrasion, planing Softening, undercutting	Wave and current flow Longshore drift Storm effects	Beach sand and shingle Spits, bars, mudflats. Storm terraces
Man	Excavation, mining, quarrying, dredging Misuse and overuse (e.g. deforestation)	Surface freight Pumping, draining	Pits, quarries, tips. Coastal protection Reclamation (e.g. Polders)

The main surface agencies are **wind action, river action, ice action, sea action** and the **action of mankind.** The results of the various processes generated by these agencies fall broadly into two categories: **wearing down** (erosion and transport) and **building-up** (transport and deposition). A related summary of surface processes and their agencies and effects is given in *Table 2.1.*

2.2 FACTORS AFFECTING WEATHERING

The term **weathering** embraces a number of (mainly atmospheric) processes which cause the breakdown and disintegration of rocks. The term **erosion** is used to describe the combined effects of disintegration and removal of material. Processes involving only physical changes are usually said to produce **mechanical** weathering, while processes involving chemical decomposition produce **chemical** weathering.

The rate and extent of the weathering of a particular rock mass is governed by a number of factors:

(a) mineral composition of the rock: some minerals (e.g. quartz) are very resistant, while others (e.g. calcite, biotite) break down quickly in certain conditions.

(b) physical properties of the mineral or rock: hard crystalline rocks (e.g. granite, quartzite) are very resistant, whereas flaky rocks (e.g. shale, schist) are weaker; poorly cemented rocks (e.g. loose sands, clays) are very weak; pervious and dissolvable rocks may 'weather' below the surface (e.g. limestone caverns).

(c) climatic conditons: in dry conditions, weathering is very slow; in warm or hot humid conditions, chemical weathering will be rapid; in wet cold (freezing) conditions, mechanical weathering will be rapid.

(d) degree and length of exposure: surface rocks weather most rapidly; sub-surface weathering is often governed by the rate of groundwater movement; removal of material (e.g. cliff falls, river scouring) will accelerate weathering; weathering is often accentuated along discontinuities in the rock mass, such as joints and bedding planes; vegetation cover may keep the rocks moist and enhance chemical weathering, while lack of vegetation may enhance mechanical weathering.

2.3 MECHANICAL WEATHERING PROCESSES

FROST

When water freezes it expands by about 9%, so that ice forming in cracks and pores will exert a bursting pressure (of about 15 MN/m^2) causing flakes of rock to split away. **Frost wedging** occurs when the ice in a crack partially melts and then, with the remaining ice keeping the crack open, refreezes, thus producing further expansion until bursting takes place.

Frost action is most vigorous in high cold mountain areas where large quantities of flaky frost-shattered rock debris, known as **scree** or **talus**, may be seen lying against the lower slopes. Porous rocks (e.g. sandstone) and flaky rocks (e.g. schists) are more susceptible to frost action than are hard crystalline rocks (e.g. granite).

Frost heaving occurs in loose and poorly consolidated deposits: since ice has a lower vapour pressure than water, water molecules are attracted to the frozen surface layer where they freeze and so produce an increase in volume. Where freezing and thawing alternate on sloping ground, the increase moisture content of the thawed surface layer may cause it to become mobile and to then flow downhill – a process known as **solifluction.**

SUB-SURFACE CRYSTALLISATION

A bursting effect very similar to frost action sometimes occurs after repeated wetting and drying of rocks containing soluble salts. As the water evaporates at the surface the salt crystallises in the pores; with each wetting more salt is dissolved and with each drying more crystals form. Eventually the surface layer is pushed away by the crystal growth.

TEMPERATURE CHANGES

In dry arid regions, day-night temperature changes of as much as 50°C, repeated every day, produce expansion and contraction in the rocks. The outer layers are more greatly affected and, in the course of time, split or peel away – a process known as **exfoliation.**

Differential expansion between minerals having varying temperature coefficients may also cause disruption, e.g. quartz and feldspar in granite.

EFFECTS OF RAIN

In addition to supplying water for frost action, rain fills streams and replenishes groundwater, and provides a water environment for chemical actions. The hydraulic action of rain is sufficient in soft or loose deposits to dislodge particles and to wash material down slopes – this is known as **rainwash.**

Gulleying occurs on slopes and serves to concentrate the run-off water, forming deeper channels.

WEARING EFFECTS

Where particles of rock are rubbed along rock surfaces or brush against each other a considerable amount of mechanical wearing takes place:
ablation – wearing or rock surfaces by wind-borne particles.
attrition – wearing of particles brushing together in wind or river streams.
corrasion – downcutting and wearing effect along the bed of a river.
abrasion – surface wearing due to wind, water or ice action.

PLANTS AND ANIMALS

The presence of vegetation will produce moist conditions and so assist chemical decomposition. Plant roots often have a mechanically disruptive effect as they push into cracks and joints. Decaying plant material produces humic acid, which will assist some chemical processes.

The removal of material by burrowing animals and the quarrying and mining activities of man can lead to significant landscape change and expose new material to atmospheric weathering.

2.4 CHEMICAL WEATHERING PROCESSES

Minerals have varying degrees of stability under atmospheric or moist ground conditions. Some minerals are extremely stable and remain unchanged even in severe conditions (e.g. quartz); others may dissolve easily in water (e.g. halite or 'rock salt'), while some may be 'attacked' by other minerals in solution (e.g. calcite, orthoclase feldspar).

The rate of decomposition depends on many factors, such as the chemical composition, crystalline structure and porosity of the rock, and also on such things as degree of exposure, temperature, humidity and concentration of dissolved compounds.

The main chemical reactions involved, either singly or in combination are as follows.

SOLUTION

Water molecules are **bipolar**, that is they carry a positive electrical charge at one end and a negative charge at the other; the equation for water may be written as H_2O or as $(H)^+(OH)^-$.

Pure water is stable and yet extremely reactive, since other ions are strongly attracted to the positive or negative ends of its molecules. **Solution** is therefore a process by which water separates ions from a solid: positive ions (e.g. Na^+, Ca^{2+}, Fe^{3+}) are captured by the negative ends and negative ions (e.g. Cl^-, O^{2-}, $SO_4{}^{2-}$, $CO_3{}^{2-}$) by the positive ends. The process tends to be accelerated as the temperature rises. **Crystallisation** and **redeposition** take place when, due to cooling or evaporation, the solution becomes saturated.

Salt flats are formed by the evaporation and refilling of saline lakes and sea cut-offs. **Leaching** is a solution process whereby the cementing minerals in sedimentary rocks are washed out, greatly assisting mechanical disintegration.

OXIDATION

Atmospheric or dissolved oxygen strongly reacts with certain metals, such as iron, to form **oxides**. Quite often oxides will hydrate and expand, and will be weaker than the original mineral. The common oxides of iron formed by weathering are:

haematite Fe_2O_3	– red iron oxide
limonite $Fe_2O_3.nH_2O$	– hydrated yellow iron oxide
goethite $FeO(OH)$	– hydrated brown iron oxide

HYDRATION

Hydration is the chemical combination of water with another compound. Unlike the simple solution effect, hydration involves a definite change in chemical composition. Hydrated compounds are often weaker and occupy a greater volume. Some examples are:

anhydrite ($CaSO_4$) + water ⟶ gypsum ($CaSO_4$ $2H_2O$)

pyroxene (e.g. augite) + water ⟶ chlorite

orthoclase feldspar + water ⟶ kaolinite ('China clay') + silica + potash (in solution)

CARBONATION

As rain falls some atmospheric carbon dioxide is dissolved in it forming weak carbonic acid. This will then react with carbonate rocks, such as limestone and dolomite, as it percolates downwards as groundwater. Thus, cracks and joints are widened to form **pot-holes (sink holes, swallets)** and **caverns**, such as those at Castleton in Derbyshire and at Cheddar and Wookey Hole in Somerset.

$$\underset{\text{(limestone)}}{CaCO_3} + \underset{\text{(carbonic acid)}}{H_2O + CO_2} \underset{\text{evaporation}}{\overset{\text{solution}}{\rightleftharpoons}} \underset{\text{(calcium bicarbonate)}}{Ca(HCO_3)_2}$$

The calcium bicarbonate solution is unstable and, as evaporation takes place, calcium carbonate is redeposited as **tufa. Stalactites** and **stalagmites** are tufa deposits formed in caves as dripping water slowly evaporates; the scale found in hot-water pipes and kettles is formed in the same way.

2.5 FORMATION OF SOILS AND SEDIMENTS

Soils are assemblages, often (but not always) in layers, of the products of weathering. The nature and properties of a given soil depends on (a) the parent rock, (b) the weathering process, (c) the degree of decomposition and (d) the amount of transportation and sorting.

Residual soils are found resting on their parent rock and consist typically of two, three or four layers, or horizons, representing stages in the weathering process as shown in *Fig 2.3.* Where the topmost layer contains organic matter and is fertile it is called **topsoil,** while **subsoil** describes generally a lower less fertile layer.

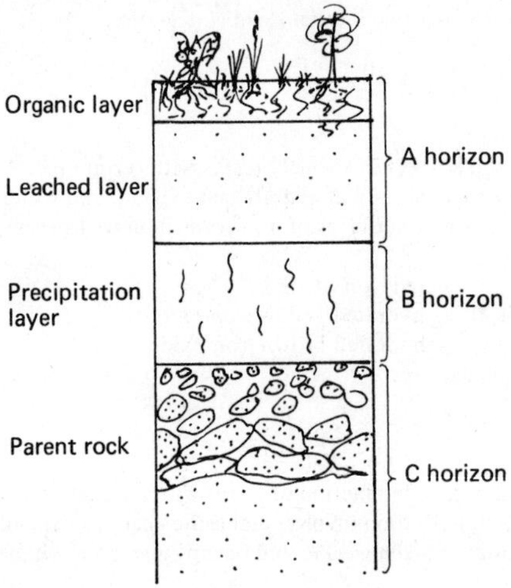

Fig 2.3 Profile of a residual soil

Transported soils have been carried to their present position by the action of gravity, wind, water or ice. The processes of transportation impose varying degrees of sorting and wear on the rock fragments.

Aeolian deposits: carried by wind action, particularly from arid regions; mainly quartz and iron oxide; well-rounded grains from sand size down to fine dust called **loess.**

Alluvial deposits: carried by streams and rivers; sub-angular to sub-rounded grains from boulder size to silt, deposited in **flood plains** and **terraces**; also include **estuarine** and **deltaic** silts and sands.

Glacial deposits: carried by ice and deposited during melting; often referred to as **drift**; **moraine gravels** are found in ridges and terraces, **boulder clay** is found in valley bottoms and in flatter areas.

Rock debris: carried by gravity; often frost-shattered fragments lying at the foot of mountains and hills; angular to flaky grains referred to as **scree** or **talus.**

Littoral deposits: shore-zone deposits, often well-sorted; rounded grains; **beach sands** and **gravels, blown sand** and shingle **spits** and **bars.**

2.6 PROCESSES OF EROSION, TRANSPORTATION AND DEPOSITION

The geological processes at work on the Earth's surface are responsible for the **erosion, transportation** and **redeposition** of crustal material. These three aspects of work may be seen in the action of **five** main agencies: **gravity, wind, rivers, ice** and the **sea**, and on a smaller, but still significant, scale, **man.**

GRAVITY

There are three main categories of downslope movement:

(a) *Landslips:* movement of relatively intact masses along definite planes, such as joints and bedding planes as in **falls**, or cylindrical surfaces in **rotational slips.**

(b) *Slides and flows:* slow (usually) and more mobile movement with mass disturbance; e.g. **mudflows,** with high moisture content, or **debris avalanches** when dry.

(c) *Creep and bulging:* **hillside creep** is a slow partial movement down a gentle slope characterised by the bending of trees and the formation of **terracettes; cambering** occurs when a clay underlying a pervious rock, such as limestone, softens and bulges, thus letting the limestone down towards the valley.

WIND ACTION

Wind is the chief erosive agent in arid regions where vegetation is sparse. **Abrasion** or **ablation** of rock surfaces is caused by wind-borne particles, which themselves become well-rounded due to **attrition**. Differential wearing of hard and soft layers produces

Fig 2.4 Stack or pedestal

stacks or **pedestals** (*Fig 2.4*). **Loess** is a wind-blown deposit often found hundreds of miles from its desert origin.

RIVER ACTION

The erosive work of a river consists of:

Abrasion	– Mechanical wearing of the river bed by water-borne particles.
Solution	– Chemical decomposition of minerals and/or cementing agents.
Hydraulic scouring	– Plucking and gouging where flow velocity is high, e.g. at bends, around bridge piers, etc.
Attrition	– Wearing and rounding of particles themselves, as they are rolled and bumped along.

Three developmental stages may be recognised in most rivers: **youthful, mature** and **senile.** Each stage is characterised by variations in both physical appearance and the

TABLE 2.2 Stages in river development

Characteristic	*Youthful river*	*Mature river*	*Senile river*
Valley	steep, V-shaped	flat	wide and flat flood plain
Gradient	steep	moderate	gradual
Velocity of flow	swift	moderate	slow and lazy
Bed	boulder strewn, with potholes	gravel to sand – shallow	not seen – deep
Course	straight, with rapids and falls	beginning to meander	wide meander loops and oxbows
Erosion	deepening	beginning to widen	widening and reworking alluvium
Deposits	boulders	gravels and sands	flood plain and terrace sands

work being done, as summarised in *Table 2.2.* The rates of erosion, transportation and deposition depend to some extent on the size (volume) of the river and the nature of the strata along its bed, but the main factor is the flow velocity (if the velocity is doubled the carrying capacity may be increased by 50–60 times).

A river carries its load in three ways: in **solution** (dissolved material), in **suspension** (the suspension load increases with flow velocity) and by **saltation** (a process of being rolled and bumped along the river bed).

Towards its mouth, a river becomes more mature and is often in the senile stage, flowing slowly in a wide valley. The deposition of **alluvium** begins in the mature stage with cobbles and gravels, and as the velocity drops in the lower reaches sands and then silts are dropped. In flood, the water may spill over the banks forming a **flood plain** or **alluvial flat.**

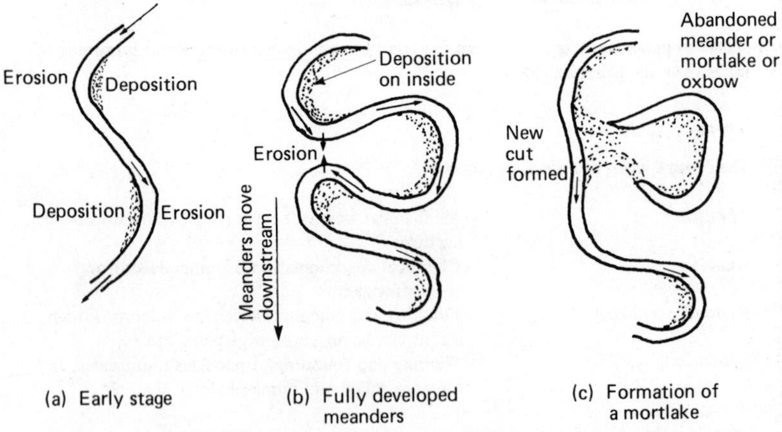

Fig 2.5 Development of meanders

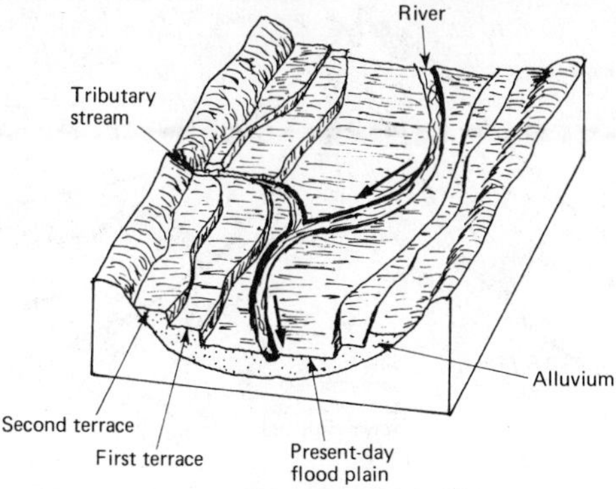

Fig 2.6 River terraces

A senile river often swings from side to side, recutting its path through the flood lain in a series of snaking loops called **meanders.** *Fig 2.5* shows how a meander may levelop and eventually become abandoned to form a **mortlake** or **ox-bow.**

If there is a fall in the estuary level, the river becomes **rejuvenated** and begins again o cut downwards into the flood plain deposits. As a new flood plain is formed, emnants of the earlier deposits may be left at higher levels along the valley sides; these re called **river terraces** (*Fig 2.6*).

CE ACTION

The accumulation of ice in upland areas causes glacial flow towards lower ground. Soil nd loose deposits are swept away and the underlying rocks worn down by the ouging, planing, scratching action of a mixture of ice and rock material. Moving **ice heets** will smooth and flatten hills and low peaks, producing (after withdrawal of ice) gently rounded area of relatively even elevation.

Valley glaciers, also loaded with rock debris, vigorously wear downwards, producing ypically over-deepened U-shaped valleys. Some of the erosional features associated vith valley glaciation are illustrated in *Fig 2.7.*

Ianging valley: glaciated tributary valley, not deepened to the same extent as the main alley.
Corrie (also *Cirque*): armchair-shaped depression formed at the head of a valley glacier; often containing a *Corrie Lake* or *Tarn.*
Arete: knife-edged ridge formed as two corries have cut back towards each other.
Horn: or *pyramidal peak,* produced by the junction of several corries.
Roche Moutonnée: a projecting hard bed in the valley floor which had been smoothed n the upstream side by glacial abrasion and plucked into a rough surface on the down-tream side due to refreezing.

As an ice sheet or valley glacier melts and 'retreats' it leaves behind a variety of leposits:

Moraine: material carried and then deposited by a glacier: *ground moraine* is carried long the base; *lateral moraine* is carried along the valley sides; *terminal moraine* is eleased from the leading edge as it melts.

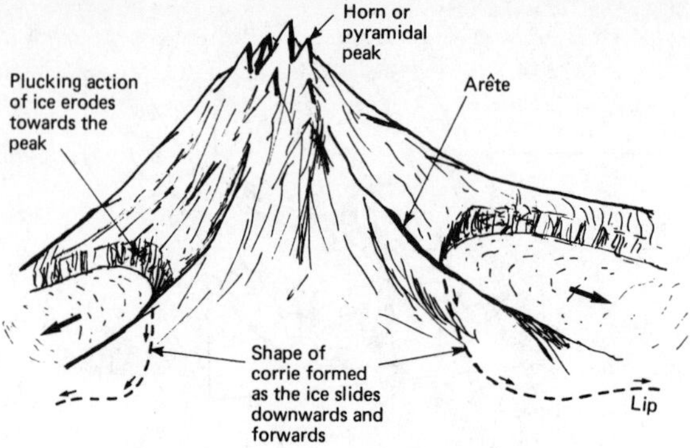

(a) Features at the head of a valley glacier

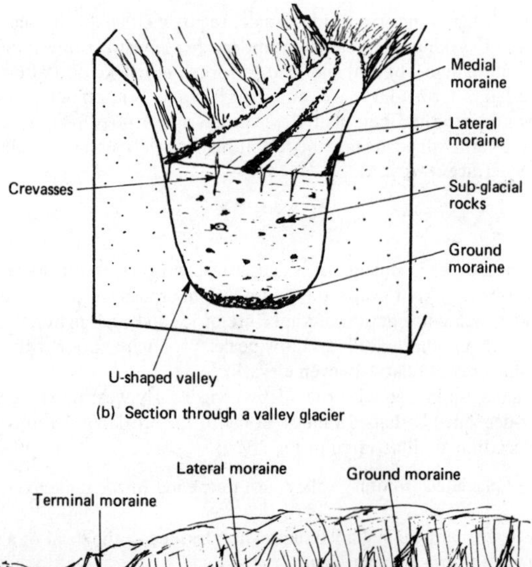

(b) Section through a valley glacier

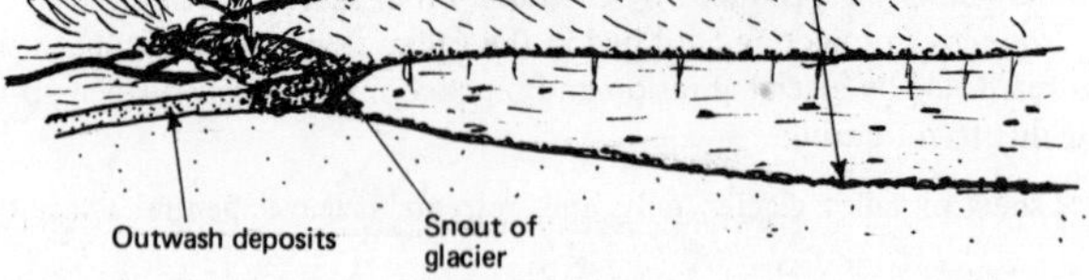

(c) Deposition at snout of a glacier

Fig 2.7 Glaciation

Boulder clay (or *Till*): unsorted glacial deposits from the main body of ice ranging from large boulders down to fine rock flour and clay.
Drumlins: elongated whale-backed mounds of boulder clay, generally occurring in groups.
Eskers: long sinuous ridges marking the previous location of sub-glacial streams.
Kames: hummocky mounds of terminal moraine.
Crag and Tail: formed by the presence of a hard bed in the valley floor, downstream of which a long stretched-out tail of moraine has been deposited.

SEA ACTION

The sea erodes the land along a coast by the composite action of waves, tides, currents and wind forces on a mixture of water and rock material. The mechanical processes include explosive expansion in joints and caves due to water and air pressure, the hydraulic wearing (jet effect) of moving water, abrasion due to rock material being dragged across the shore zone rocks and the softening of some cementing agents.

Hard rocks tend to be weakened along joints and bedding planes, so that from time to time falls of rocks occur leaving new steep or even vertical cliff faces. Soft rocks

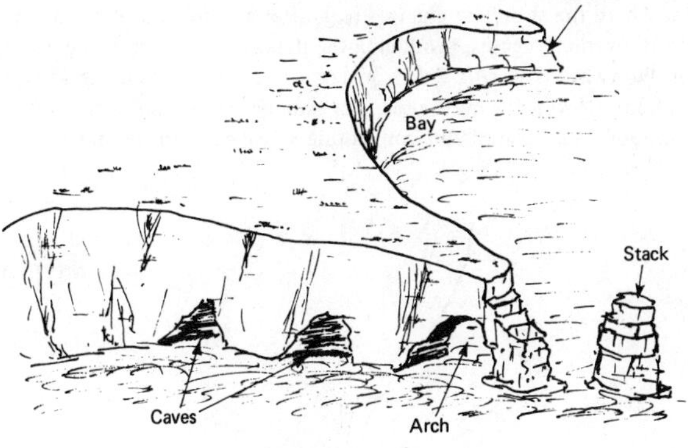

Fig 2.8 Coastal cliff features

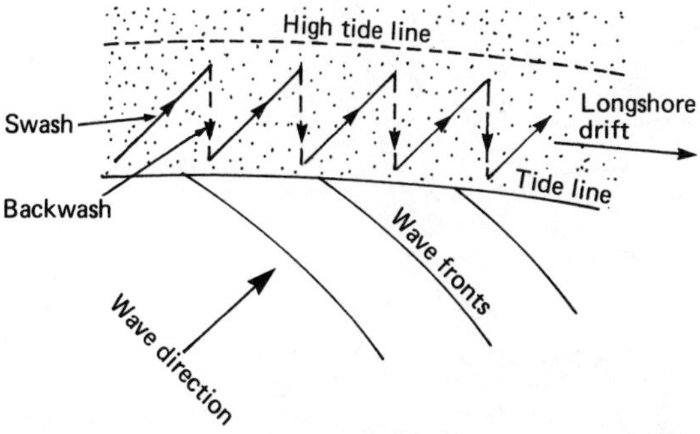

Fig 2.9 Longshore drift

such as clays and shales weather back to a shallower angle, and may degrade further d to rotational landslipping.

Caves are formed due to joint widening (*Fig 2.8*). At headlands and promontories persistant erosion in cave openings may result in **arches**, following the collapse of which **coastal stacks** will be left standing clear of the cliff line. (e.g. Needles, I.O.W.; Old Harry Rock, Dorset; Old Man of Hoy, Orkneys).

Between the cliff-line or high-water mark and the low-water mark the twice daily continuous tidal movement creates a planing action which produces a flat **wave-cut platform**. Along sections of coast where deposition can occur, the wave-cut platform may be covered with beach **sand** or **shingle**, but in scoured areas the wave-cut platform is exposed. In storm conditions, unusually large amounts of beach material may be scoured away, or alternatively carried high up the beach and deposited as **storm terraces**.

When the general trend of movement due to wind, current and tide is oblique to the shoreline, shingle is moved along the coast by a process termed **longshore drift**. The movement of waves up the beach (**swash**) carries particles obliquely towards the shore (*Fig 2.9*). The **backwash**, however, falls back along the steepest slope (i.e. perpendicular to the shoreline). The net effect is therefore to move particles **along** the shore.

At a point where the shoreline changes (e.g. after a headland, or at a river estuary) the longshore movement ceases, producing the deposition of a **spit** (e.g. Dungeness, Spurn Point, Westward Ho,. (*Fig 2.10*). Where a spit continues across a bay or lagoon, cutting off a body of water from the sea, a coastal **bar** is formed (e.g. Slapton Sands, Devon). A **tombolo** is a form of spit connecting an island with the mainland (e.g. Chesil Beach, Dorset).

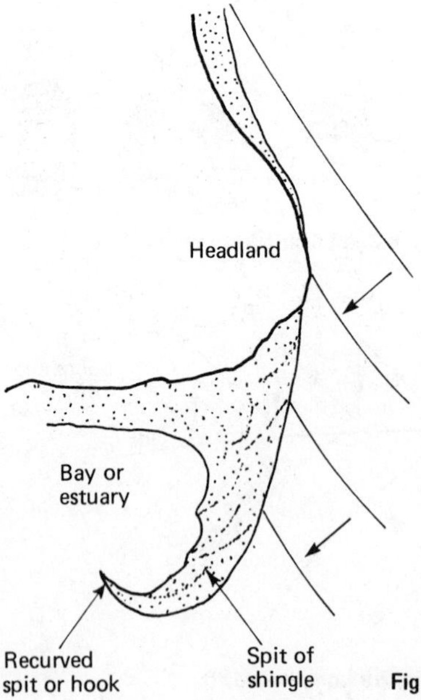

Fig 2.10 Formation of a spit

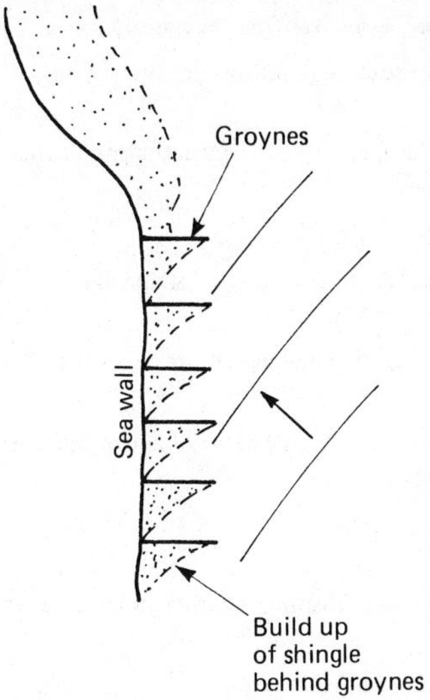

Fig 2.11 Groynes

Protection is afforded to sections of coastline by the provision of **sea-walls** and **harbour walls** designed to resist heavy wave action, and **groynes** designed to limit longshore drift (*Fig 2.11*).

ACTION OF MAN

Large quantities of material from the Earth's crust are excavated or mined by man and transported considerable distances. The spoil from industrial processes is piled up high in tips, or used to reclaim poor land. Rivers are dredged, sea-walls are built, cuttings and tunnels are driven through mountains and deep holes are drilled in the ground. Although man's Earth-shaping endeavours may appear somewhat puny alongside some of the spectacular events provided by nature, they are in many cases significant enough, and even a cause for concern.

The Earth's mineral resources have been created over millions of years, but the rate at which mankind is using some of these is alarmingly fast. At one opencast mine in Australia (Mount Tom Price) about 60 000 tonnes of iron ore is being removed every day. The known deposits of copper (essential in electrical/electronic components) are almost exhausted; these may not last another 100 years. The rate of mineral withdrawal from the crust is estimated to be about one-third of the rate at which new crust is being formed at constructive plate margins. The mineral budget of the Earth's crust is in grave danger of being overspent.

TEST EXERCISES

COMPLETION QUESTIONS (answers on page 204)

Complete the following statements by inserting an appropriate word or words in the spaces indicated.

1 The surface of the Earth is continually . . . .

2 The cyclic nature of geological processes was first recognised by . . . .

3 The five main agencies involved in surface processes are the actions of . . . .
. . . .

4 The factors which affect the rate and extent of the weathering of surface rocks are: . . . .

5 When water freezes it expands by . . . .

6 Frost-shattered debris has a characteristically . . . . . . shape and is known as . . . . or . . . .

7 The process of alternate freezing and thawing which produces slope movement is called . . . .

8 The process of . . . . involves the peeling away of successive surface layers of rock due to temperature changes.

9 The general wearing of rock surfaces is termed . . . . , whereas . . . . refers only to the action of wind.

10 The wearing and rounding of particles brushing together in wind or water streams is referred to as . . . .

11 List four chemical processes involved in weathering . . . .

12 Water consists of positive and negative ions, which are . . . .

13 The process whereby water separates ions from a solid is called . . . .

14 Leaching is a . . . . process, wherein the . . . . mineral in a rock may be removed.

15 State the names of three common oxides of iron . . . .

16 When water chemically combines with another compound, forming a new mineral, ... is said to take place. This process is involved in the formation of both kaolinite from . . . . feldspar and chlorite from . . . .

17 'China clay' is a common name referring to . . . .

18 The process of . . . . involves the reaction between atmospheric carbon dioxide and limestone rock. Give the equation for this reaction . . . .

19 The process described in Question 18 causes the widening of joints and cracks to form . . . .

20 Stalactites and stalagmites are forms of redeposited . . . . , referred to collectively as . . . .

21 Non-transported soils are referred to as . . . . soils.

22 Insert the names given to groups of deposits according to their transportation history: carried by wind: . . . . ; carried by ice . . . . ; carried by rivers . . . . ; carried by gravity . . . .; carried by sea along the shore . . . .

23 The three main categories of downslope movement are: . . . .

24 The three developmental stages seen in most rivers are: . . . .

25 A river carries its load in three ways, namely in . . . . , in . . . . and by . . . .

26 The wide flat area adjacent to the lower reaches of a river is called a . . . .

27 The snaking loops seen in the courses of some rivers are called . . . . , when cut-off and abandoned these form . . . .

28 State four erosional features evident in a glaciated valley: . . . .

29 Material deposited by glaciers is termed . . . . Identify the following types:
Unsorted large masses or sheets: . . . .
Elongated whale-backed ridges: . . . .
Long sinuous ridges: . . . .
Hummocky mounds: . . . .

30 Erosion along a coast produces a planing action resulting often in the formation of a . . . .

31 The movement of beach material parallel to the coast is termed . . . . Three examples of depositional formations resulting from this are: . . . .

ESSAY QUESTIONS

1 Discuss the notion of continual change and the concept of a geological cycle. Why are these ideas important and fundamental to the understanding of geological processes?

2 Give an illustrated account of the processes involved in the weathering and disintegration of rocks.

3 List the various physical and chemical processes that work to bring about the wear and disintegration of exposed rock surfaces and discuss these mechanisms in the context of choosing suitable natural building materials.

4 Draw up a classification of natural soils according to their depositonal and transportational history. Discuss the principal engineering problems associated with each class.

5 Give an illustrated account of the various processes associated with downslope movement and instability.

6 Give an illustrated account of the work of rivers, including erosional, transportational and depositional aspects.

7 Using annotated sketches, explain the formation and appearance of the features associated with valley glaciation.

8 Give an illustrated account of the processes and features associated with coast erosion, commenting on any related engineering problems.

9 Discuss the role of mankind in the context of geological changes.

3 Common rocks and minerals

3.1 ELEMENTS, IONS AND MINERALS

Rocks are made up of different **minerals**, sometimes in crystaline form, sometimes as mineral grains bound together with mineral cement. There are about 92 natural elements, but only a very small number of these feature prominently in the composition of crustal rocks. In fact, nearly three-quarters of the Earth's crust is made up of just two elements: oxygen and silicon. As shown in *Table 3.1,* these two non-metals, together with six metallic elements, account for almost all crustal rock: over 98% by weight and over 99.9% by volume.

TABLE 3.1 Occurrence of natural elements

Element		*Proportion in whole earth (%)*	*Proportion in crust*	
Name	*Symbol*		*By weight (%)*	*By volume (%)*
Oxygen	O	30	46.6	93.8
Silicon	Si	15	27.7	0.9
Aluminium	Al	1.1	8.1	0.5
Iron	Fe	35	5.0	0.4
Calcium	Ca	1.1	3.6	1.0
Sodium	Na	incl. in 'others'	2.8	1.3
Potassium	K	incl. in 'others'	2.6	1.8
Magnesium	Mg	13	2.1	0.3
Nickel	Ni	2.4	incl. in 'others'	
Sulphur	S	1.9	incl. in 'others'	
Others		0.5	1.5	< 0.1

Elements having a tendency to **lose** their outermost electrons are known as **metals** and form positively-charged **metallic ions**, e.g. Na^+, K^+, Mg^{2+}, Ca^{2+}, Fe^{3+}. Elements which tend to **gain** electrons are known as **non-metals** and form negatively-charged ions, e.g. F^-, Cl^-, O^{2-}, S^{2-}, N^{3-}, P^{3-}. Many mineral compounds consist of **ionic** combinations of metallic and non-metallic ions, e.g. halite (rock salt) NaCl, hematite (iron oxide) Fe_2O_3. Certain combinations of ions form non-metallic groups, which also combine with metallic ions to form minerals, e.g. calcite $CaCO_3$, gypsum $CaSO_4 2H_2O$.

Some elements are neither completely metallic nor non-metallic, but possess instead a capability for sharing electrons. The simplest electron sharing, or **covalent** structure, is that of **diamond**, in which each **carbon** atom shares electrons with four other carbon atoms spread around it in a tetrahedral framework. **Organic** compounds are based on complex covalent chains and rings of carbon atoms, giving rise to the vast field of organic chemistry.

Silicon also is an element capable of covalently combining to form chains, sheets and frameworks. This it does by sharing electrons with oxygen ions, which in turn may share electrons with other oxygen ions or other metallic ions. Thus, a family of silicon-oxygen compounds exists known as the **silicates**. This is the predominant group of rock-forming minerals. Other major groups of rock-forming minerals are **oxides, sulphides, halides, carbonates, sulphates** and **phosphates**.

3.2 IDENTIFICATION OF MINERALS

A **mineral** may be defined as an inorganic substance having a definite chemical composition and a characteristic crystalline structure. A **rock**, on the other hand, is an aggregate of one or more minerals, often varying in content and distribution.

Although several thousand mineral varieties exist, most of them are very rare, and only a relatively small number may properly be described as **common.** In order to assess the properties of rocks, it is necessary first to identify them by identifying their predominant minerals. Identification techniques fall into a number of categories:

(a) *Examination of hand specimen:* a study of the physical characteristics of a hand-held specimen of rock or mineral, using the naked eye or a X10 hand lens.

(b) *Optical examination of a thin slice:* a thin slice of mineral or rock is ground down to a thickness of 30 μm (at which many minerals are transparent) and mounted on a glass slide; examination for optical characteristics is carried out using a petrological microscope.

(c) *Chemical analyses:* several types of chemical analyses, including spectographic and thermal dissociation techniques, can be used both qualitatively and quantitatively.

(d) *X-ray and other irradiation methods:* observations of the scattering of X-rays, etc., can yield details of crystalline structure. For the purpose of the *Geotechnics 4* unit only the examination of hand specimens will be dealt with. For details of other techniques, students should refer to suitable texts on geology or mineralogy.

PHYSICAL CHARACTERISTICS OF MINERALS

Certain of the physical characteristics displayed by minerals provide a means of identifying hand specimens. *Table 3.2* gives a summary of the most commonly used

TABLE 3.2 Physical characteristics of minerals

COLOUR: colour of freshly broken surface

green – chlorite, olivine
dark green – hornblende
brown – biotite red – cinnabar
yellow – sulphur grey – galena

STREAK: colour of powder – shown by the mark left on a piece of unglazed porcelain.

red – haematite
grey – magnetite brown – limonite
yellow – realgar black – galena

FORM: appearance of a crystalline mass or cluster (not the crystal-structure shape).

single crystals
prismatic (parallel faces):
cubic – galena
hexagonal – quartz
rhombohedral – calcite
tabular (flat or platy) – gypsum
acicular (needle-like) – tourmaline

clusters or masses
banded – agate
dendritic (branching) – manganese dioxide
drusy (lining a cavity) – amethyst
fibrous – asbestos
granular – calcite, quartz, magnetite
reniform (kidney-shaped) – haematite
nodular (irregularly rounded) – flint

HARDNESS: comparative hardness (scratchability) according to *Moh's scale of hardness*

1. Talc
2. Gypsum – – – – finger nail
3. Calcite – – – – copper coin
4. Fluorspar
5. Apatite – – – – knife blade, window glass
6. Orthoclase
7. Quartz – – – – steel file
8. Topaz
9. Corundum
10. Diamond

Note: the divisions are not equal; diamond is the hardest known natural mineral and is several times harder than corundum.

LUSTRE: appearance of surface in reflected light

metallic – galena, pyrite
vitreous (glassy) – quartz
pearly – muscovite
resinous – sulphur
silky – asbestos, satin spar
dull – flint, wad

CLEAVAGE: tendency to split easily along definite *cleavage planes* in 1, 2 or 3 directions.

perfect X 1D – micas
perfect X 2D – hornblende
perfect X 3D – galena, calcite
good X 2D – orthoclase
poor X 2D – olivine
indistinct – pyrite
none – opal, quartz

FRACTURE: nature of broken surface (independent of cleavage)

conchoidal (shell-like) – quartz, flint, glass
even (nearly flat) – chert, jet
uneven (rough) – tourmaline (many others)
hackly (spiky) – native silver

OPTICAL EFFECTS:

transparency: transparent – quartz
translucent – gypsum
opaque – most minerals.
iridescence – chalcopyrite
play of colours – labradorite
chatoyancy – crocodilite
fluorescence (in UV light) – fluorspar
double refraction – calcite

SPECIFIC GRAVITY: heaviness compared with water

Range: 1 to 22 (Platinum = 21.45)
Average: 2.6
Quartz, calcite, feldspar $\doteq$ 2.7
Metallic oxides & sulphides 4.0–8.0

OTHERS: chemical composition, fusibility, tenacity, magnetism, feel, smell, taste, radioactivity.

characteristics, together with typical examples. While most minerals possess one or two obvious characteristics, not all of the features will be displayed by a single mineral; the same characteristic may also be evident in several different minerals. For example, not all minerals exhibit cleavage; also, many minerals are white in colour.

In the majority of cases, the common minerals can be identified from two or three of these characteristics – providing the observer has a trained eye and some experience.

3.3 SOME COMMON MINERALS

SILICATES

The most abundant and important minerals in the Earth's crust are **silicates**, and these are all based on the tetrahedral combination of silicon and oxygen shown in *Fig 3.1*. The $SiO_4{}^{2-}$ group has the ability to share one or more of its oxygens with other similar

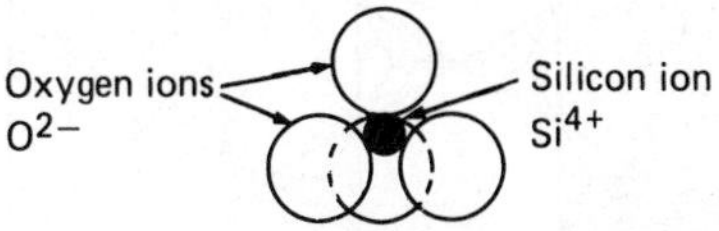

Fig 3.1 Basic tetrahedral unit of silica

tetrahedral units and so form a series of different structures. The silicate units (negatively-charged) can be linked together by positively-charged metallic ions (K^+, Mg^{2+}, Ca^{2+}, Fe^{2+}, etc.) giving rise to a wide range of combinations.

In *Table 3.3*, some of the most common examples of silicate structures are shown.

Light-coloured silicates

Quartz is a three-dimensional framework of silica tetrahedra having a unit structure of SiO_2. In a pure form it is colourless and glassy, and typically forms six-sided crystals. Traces of impurity give rise to coloured varieties: **rose quartz** (pink), **smoky quartz** (grey-brown), **amethyst** (purple, **milky quartz** (white).

Quartz is a primary constituent of acid igneous rocks and is also common in sandstones, quartzites and some metamorphic rocks. **Chert** is a cryptocrystalline variety of silica (SiO_2) occurring in bands in sedimentary rocks, or in nodules as **flint**; **jasper** is a red variety of chert.

Chalcedony is cryptocrystalline silica having a fibrous or banded texture; varieties include **agate, onyx** and **carnelian. Opal** is a hydrated form (SiO_2 nH_2O) of chalcedony.

The **feldspars** are also frameworks of silica, but with some Al substituted for Si, and with tetrahedra linked with K^+, Na^+, or Ca^+ ions.

potash feldspar – orthoclase	$Ka\ Al\ Si_3\ O_8$	
sodic feldspar – albite	$Na\ Al\ Si_3\ O_8$	Plagioclase family
calcic feldspar – anorthite	$Ca\ Al\ Si_2\ O_8$	

Feldspars are important constituents of many igneous rocks, orthoclase and albite being common in acid rocks (e.g. granite, syenite) and anorthite being found in basic rocks (e.g. gabbro).

TABLE 3.3 Silicate mineral structures

Combination	*Structural unit*	*Common examples*
Single tetrahedra	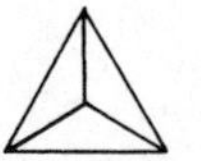	Olivine
Double tetrahedra	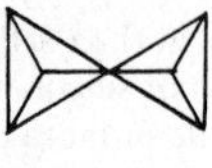	Melilite
Closed ring	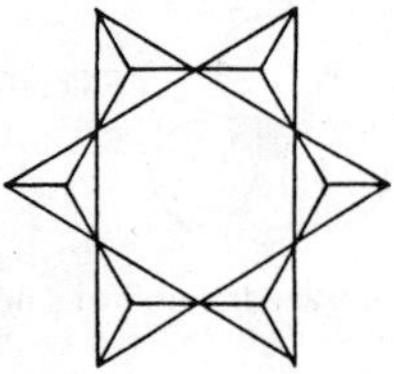	Beryl (n = 6)
Single chain	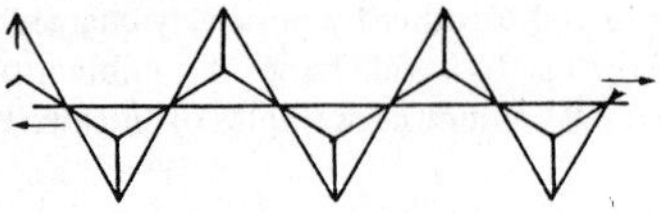	Pyroxenes (e.g. augite)
Double chain	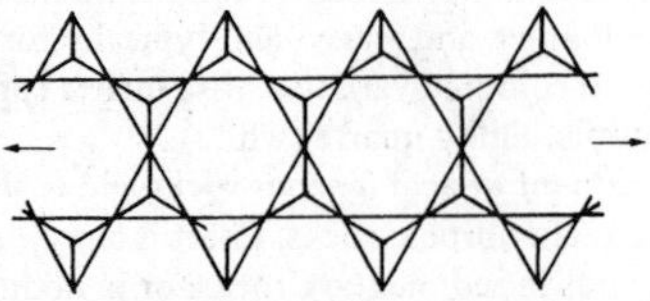	Amphiboles (e.g. hornblende)
Sheets	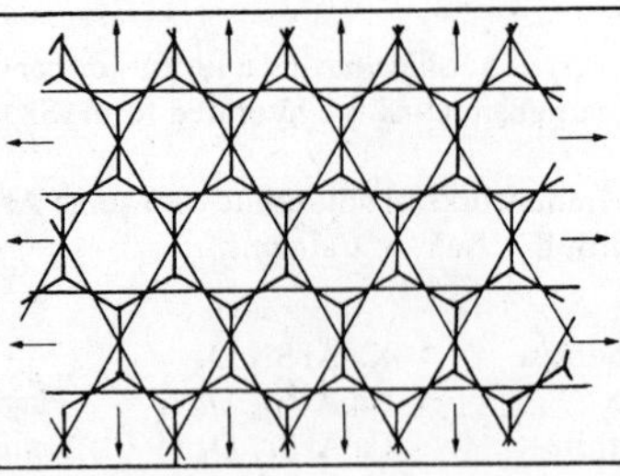	Micas (e.g. muscovite, biotite)
3D frameworks		Quartz Feldspars

Clay minerals are formed by the decomposition of silicate minerals, chiefly orthoclase and plagioclase, as follows:

orthoclase + water $\longrightarrow$ kaolin + silica + potash (KOH)
(china clay – no K^+ left)

or orthoclase + water $\longrightarrow$ illite + silica + potash (KOH)
(clay – some K^+ left) (incomplete)

and plagioclase + water $\longrightarrow$ montmorillonite + calcium hydroxide
(fuller's earth clay)

Dark-coloured or ferromagnesian silicates
Those silicates containing essential Mg or Fe are usually dark-coloured and are referred to as **ferromagnesian** minerals.

Olivine is a (Mg, Fe) silicate consisting of separate tetrahedra linked by Mg and Fe ions in varying proportions. Olivine commonly occurs in basic and ultrabasic rocks (e.g. basalt and picrite) and is probably the main constituent of the Earth's mantle. The alteration of olivine leads to the formation of **serpentine**:

olivine + water + $CO_2 \longrightarrow$ serpentine + magnesite ($MgCO_3$)

Pyroxenes are a group of minerals having linked chains of tetrahedra, the most common of which are:

Enstatite $Mg\,Si\,O_3$, Hypersthene $(Mg, Fe)\,Si\,O_3$
Augite $Ca\,(Mg, Fe, Al)(Si, Al)_2\,O_6$, Diopside $(Ca, Mg)Si_2\,O_6$

The **Amphiboles** are a large group of minerals comprising linked double chains of tetrahedra. The most common amphibole is **hornblende,** a complex alumino-silicate of (Ca, Mg, Fe, Na, Al) occurring in many igneous rocks (e.g. syenites, diorites, andesites) and in some metamorphic rocks (e.g. hornblende schist). Certain fibrous varieties of amphibole, i.e. **tremolite** and **crocodilite**, are known as forms of **asbestos**. Another form of asbestos is **chrysotile**, a fibrous type of serpentine.

The **Micas** are a group of characteristically flaky minerals, made up of pairs of silica sheets linked by Fe, Mg, K or OH ions.

Muscovite: light-coloured or transparent due to absence of Mg and Fe; occurs in acid igneous rocks, schists, gneisses, and sediments.

Biotite: 'brown mica'; brown to black due to presence of Mg and Fe; occurs in many igneous rocks; alters to form clays such as **illite** and **vermiculite.**

Chlorites are green flaky minerals similar to the micas, but containing much more OH. They occur chiefly as the alteration products of ferromagnesian minerals, such as biotite, augite and hornblende, in basic igneous rocks and some metamorphic rocks.

OXIDES

Since oxygen is both abundant and highly reactive, it appears in most minerals; however, when a direct metal-oxygen compound is formed it is called an **oxide**. The most common are:

Haematite (Fe_2O_3): red ferric oxide, occurring widely in different forms, such as crystalline nodules and veins, as an earthy deposit **(red ochre)** and as a cementing agent in sediments; it is the most important ore of iron.

Limonite or Goethite: hydrated iron oxide; brown to yellow in colour; occurring often with haematite and as an earthy deposit (yellow ochre).

Magnetite Fe_3O_4: magnetic iron oxide; occurring as grains or crystals in igneous rocks and sediments; important ore of iron when it has been concentrated by gravimetric separation.

Ilmenite	$FeTiO_3$	iron-titanium oxide
Cassiterite	SnO_2	ore of tin
Cuprite	Cu_2O	an ore of copper.

Bauxite: this is the primary ore of aluminium and comprises two hydrated aluminium oxides, **gibbsite** $Al_2O_3\,3H_2O$ and **diaspore** $Al_2O_3\,H_2O$; occurring as earthy masses or concretions, in association with clay and often with iron oxide and silica.

SULPHIDES

Pyrite FeS_2: the most abundant sulphide mineral; occurring in igneous rocks, hydrothermal veins, shales, slates and schists: yellow brassy colour (hence 'fool's gold'), often in the form of cubic or pyritohedral (12-sided) crystals.

Galena PbS: an important ore of lead; forms perfect cubic crystals; usually occurring in veins and lodes, associated with calcite or quartz or other metallic minerals.

Sphalerite (zinc blende) ZnS: the main ore of zinc; occurring in veins, metamorphic rocks, often with galena; forms tetrahedral crystals.

Chalcocite Cu_2S and **Chalcopyrite** $CuFeS_2$: are ores of copper: occurring in veins and lodes, often with other sulphides.

CARBONATES

The carbonates occurring as primary sedimentary minerals are **calcite, dolomite** and **siderite.**

Calcite $CaCO_3$: characterised by strong 3D rhombohedral cleavage and dissolves in weak acids; often forms 'dog-tooth spar' crystals in druses; predominant mineral in limestones, also in veins and as a secondary mineral in basic igneous rocks.

Dolomite $CaMg(CO_3)_2$: double carbonate, also rhombohedral crystals; not so soluble in cold dilute acid; predominant mineral of dolomite rock; also in dolomitic limestone.

Siderite $FeCO_3$: rhombohedral crystals; important type of iron ore.

Magnesite $MgCO_3$: alteration product of magnesian silicates in igneous rocks.

EVAPORITES

Evaporites result from the evaporation of saline water, e.g. sea cut-offs, saline lakes. The least soluble minerals are precipitated first, followed by others in order of increasing solubility.

Sylvite	potassium chloride
Carnallite	potassium magnesium chloride
Halite	sodium chloride (common rock salt)
Epsomite	magnesium sulphate
Glauberite	calcium-sodium sulphate.

Gypsum $CaSO_4\ 2H_2O$: **selenite** is a tabular or diamond-shaped crystalline variety; **alabaster** is a white or pink massive form; in **satin spar** the crystal structure is acicular (needle-like); occurring as an evaporite, also due to decomposition of pyrite; found in many clays and limestones. **Anhydrite** is a anhydrous evaporitic form ($CaSO_4$).

FLUORINE MINERALS

Fluorspar (fluorite) CaF_2: perfect cubic cleavage gives tetrahedral crystals; occurring in veins, often with other sulphides; **Blue-John** is a purple variety from Derbyshire used for decorative purposes.
Apatite $Ca_5F(PO_4)$: this is a phosphate; occurring in igneous rocks and veins.

3.4 NATURE AND CLASSIFICATION OF ROCKS

Any naturally occurring coherent assemblage of minerals is a **rock**. Rocks are solid (but not necessarily hard) parts of the Earth's crust; **soils** on the other hand are loose unconsolidated deposits. The majority of rocks are made up of mineral mixtures; a few contain organic material (e.g. coal, phosphorite). Some, although being entirely mineral in composition, have biogenic origins, e.g. chalk and some other limestones.

The main basis for classifying rocks is their origin or mode of formation, with their mineral content often being used for cross classification. Three major groups are defined.

Igneous rocks are formed by cooling from a molten state and are therefore **holo-crystalline** (entirely crystalline); the slower the cooling process, the larger the crystals.

Sedimentary rocks are formed from deposited sediments, which have subsequently been consolidated and cemented together to produce coherent layers **(strata)**. Sediments consist variously of weathered and transported debris eroded from land masses, organic and skeletal material from animal and plant life and minerals precipitated from bodies of water.

Metamorphic rocks are altered rocks; alteration may be due to contact or proximity to igneous intrusions or to the effects of pressure and temperature changes brought about by tectonic movements.

3.5 IGNEOUS ROCKS

Molten rock material, called **magma**, is formed in the Earth's crust at interactive plate margins, mainly due to high pressure and the generation of frictional heat. As a mobile liquid under high pressure, magma is forced along lines of weakness, such as faults, joints, bedding planes, etc. In some places it is extruded on to the surface, through **volcanic vents**, forming **lava flows** (*Fig 3.2*). A lot of magma, however, fails to reach

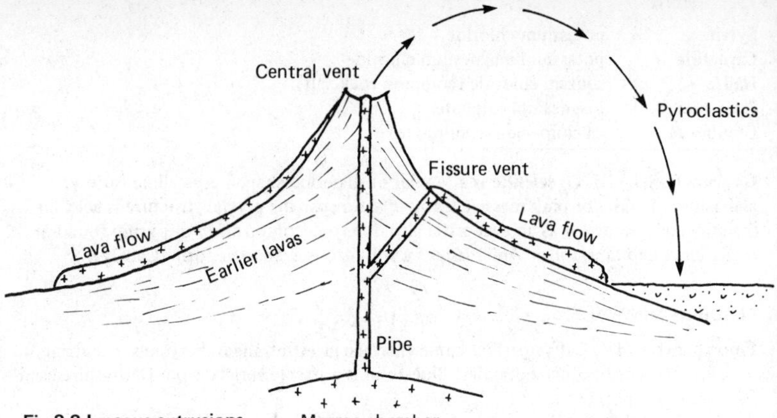

Fig 3.2 Igneous extrusions

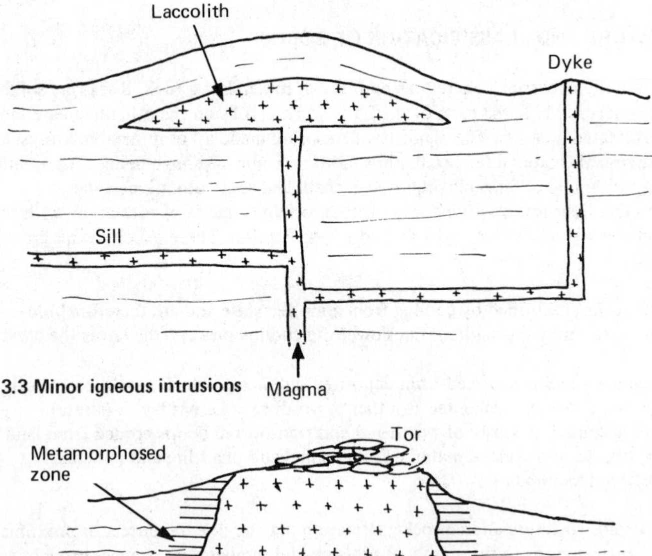

Fig 3.3 Minor igneous intrusions

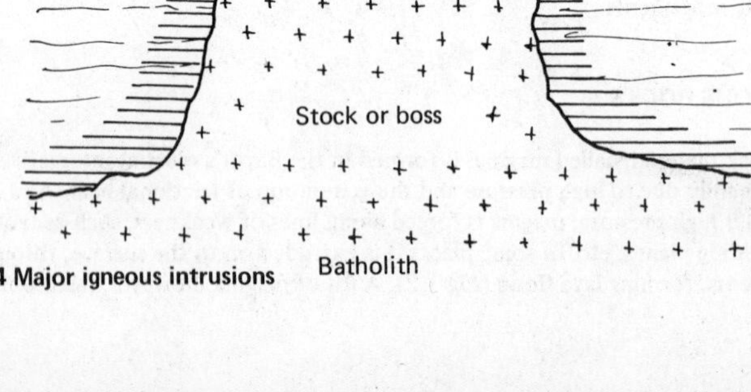

Fig 3.4 Major igneous intrusions

e surface and solidifies as it cools within the existing crustal formations as **igneous intrusions** (*Fig 3.3 and 3.4*).

The rate of cooling controls the texture of the subsequent rock. **Lavas**, which cool ry quickly on the surface, have a very fine-grained or even glassy texture. Intrusive aterials in **dykes** (vertical sheets) and **sills** (horizontal sheets) cool more slowly, roducing a fine-to medium-grained texture. Large deep-seated masses, such as **atholiths** (very large), **stocks** and **bosses** (offshoots from batholiths) cool slowly, roducing a coarse-grained texture.

Three classes of igneous rocks are defined on the basis of their mode of occurrence nd resulting texture:

xtrusive	*Volcanic* – lavas	– very fine-grained to glassy
inor intrusive:	*Hypabyssal* – sheets	– fine to medium grained
ajor intrusive:	*Plutonic* – large masses	– coarse-grained

n addition to these crystalline rocks, volcanic activity also produces a variety of non-rystalline materials collectively referred to as **pyroclastic** deposits:

uff	is consolidated and cemented volcanic ash.
gglomerate	consists of ejected fragments, usually in an ashy or cindery matrix; gravel-sized fragments up to about 30 mm are called **lapilli** and larger pieces are called **bombs.**
umice	is a highly vesicular (frothy) acid lava.

n addition to a textural/mode-of-occurrence classification, **four** compositional classes an be identified according to the mineral content of the magma: acid, intermediate, asic and ultrabasic. (See *Table 3.4*).

cid igneous rocks

cidic magmas occur mainly as deep-seated massive intrusions in continental areas. hey are silica-rich, containing > 10% quartz (free silica) and having a total silica ontent of > 66%. The essential minerals are quartz, orthoclase feldspar (often with lbite) and either muscovite or biotite (or both), with accessory minerals such as ornblende, plagioclase and apatite and with secondary minerals such as tourmaline, aolinite and topaz. Acid lavas are highly viscous and are associated often with xplosive eruptions.

porphyritic texture is often developed in which large crystals (**phenocrysts**) of a articular mineral (usually a feldspar or quartz) are set in a relatively finer ground-nass; as for example the **Shap Granite** and **Dartmoor Granites,** both of which have henocrysts of orthoclase.

ranite: very common **coarse-grained** acid rock; e.g. Cornish granite, many varieties in cotland and Ireland.

licrogranite: medium-grained equivalent; occurring in smaller intrusions (stocks, osses).

ranodiorite: similar to granite, except the plagioclase content exceeds that of rthoclase; e.g. Mountsorrel (Leics.), Hollybush (Malvern Hills), Dalbeattie (Scotland).

Texture	Grain size (mm)	Mode of occurrence	Total silica content (%): Acid (>66)	Intermediate (66–55)	Basic (55–44)	Ultrabasic (<44)
Coarse-grained Porphyritic	2	PLUTONIC (large masses)	GRANITE GRANODIORITE MICROGRANITE	SYENITE DIORITE	GABBRO NORITE	PERIDOTITE PICRITE SERPENTINE DUNITE
Medium to fine-grained Porphyritic	0.06	HYPABYSSAL (sheets)	Quartz PORPHYRY Orthoclase FELSITE	LAMPROPHYRE Plagioclase PORPHYRY	DOLERITE	
Very fine-grained Microcrystalline Glassy	0.002	VOLCANIC (lavas)	RHYOLITE DACITE OBSIDIAN	TRACHYTE ANDESITE	BASALT	

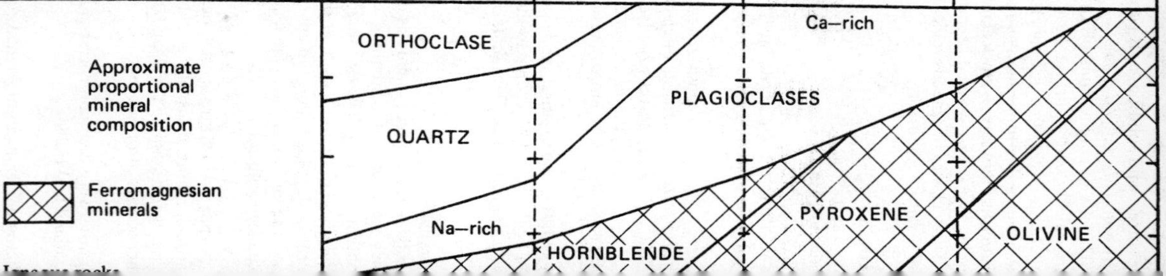

TABLE 3.4 Igneous rocks

uartz porphyry: porphyritic microgranite with phenocrysts of quartz; also called **anite porphyry** or **elvan** (in Cornwall). Also orthoclase-porphyry≡microgranite.

hyolite: very fine-grained to glassy acid lava; **pitchstone or obsidian** is a black wholly assy variety; e.g. Hebrides and Obsidian Cliff in Yellowstone Park, U.S.A.

elsite: is a devitrified rhyolite.

TERMEDIATE IGNEOUS ROCKS

termediate magmas are formed at destructive plate margins along oceanic/continental oundaries. They contain $< 10\%$ quartz, with either orthoclase, or plagioclase or both, d with accessory minerals such as mica, hornblende and sometimes pyroxene ugite).

yenite: coarse-grained intermediate rock in which the feldspar content is edominantly ($> 50\%$) orthoclase; **Larvikite** is an example from Norway containing soda-rich feldspar which displays a blue/green 'play of colours'.

iorite: coarse-grained intermediate rock in which plagioclase predominates.

amprophyre: medium-grained equivalent of syenite or diorite occurring in dykes and lls; is porphyritic with phenocrysts of orthoclase and/or plagioclase.

agioclase-porphyry (preferred to **diorite-porphyry** or **porphyrite**): medium-grained quivalent or diorite.

rachyte: very fine-grained (lava) equivalent of syenite, e.g. Central Valley of Scotland.

ndesite: very fine-grained (lava) equivalent of diorite.

ASIC IGNEOUS ROCKS

asic magmas are extruded at constructive plate margins in mid-oceanic areas, e.g. awaii, Iceland. They are rich in ferromagnesian minerals, such as olivine and yroxene, with little or no quartz or orthoclase. Secondary minerals commonly include lorite and serpentine.

abbro: coarse-grained basic rock; e.g. Cornwall, Cumberland, Lake District, berdeen, Skye.

olerite: fine to medium-grained equivalent of gabbro, occurring in dykes and sills; so called **micro-gabbro**; e.g. Midland Valley of Scotland, Whin Sill (N. Eng.), leveland Dyke (Yorks.), Clee Hills (Salop.).

asalt: the most abundant igneous rock; very fine-grained to glassy, brown or black va; may be amygdaloidal, i.e. containing almond-shaped vesicles (amygdales), in hich secondary minerals, such as chlorite and calcite, may occur. Widespread occurrences include N. Ireland, W. Scotland and Hebrides, Mendip Hills, Hawaii, Central rance, Snake River (USA) and the Deccan (India).

Ultrabasic magmas also occur in oceanic zones, but mainly as large emplacements (rather than lava flows). It is likely that ultra-basic magma emanates from the upper mantle and therefore forms new crustal rocks. Essential mineral is predominantly olivine, with augite, pyroxene and often hornblende. Little or no quartz or feldspar i present.

Picrite: typical ultrabasic rock, e.g. Cornwall, Ayrshire.

Peridotite: essentially olivine (peridot = olivine, Fr).

Dunite: almost entirely olivine.

Serpentine: produced by alteration of peridotite or dunite, i.e. olivine has altered to serpentine mineral, e.g. Lizard, Cornwall.

Chrysotile: a fibrous variety of serpentine which is a source of asbestos.

3.6 SEDIMENTARY ROCKS

Sedimentary rocks are formed from material (**detritus**) removed from pre-existing rock masses by weathering and erosion. They are usually laid down in layers, after transportation by such agencies as water, wind, ice and gravity. Deposition in large bodies of water (e.g. oceans, seas, lakes) results in extensive parallel **strata** (layers) of consistent thickness. Under continental or coastal deposition conditions the bedding is often uneven and discontinuous.

Following the deposition of loose sediments, a combined process of **consolidation** and **cementation** over a long period of time results in the formation of a coherent ma of rock. The consolidation pressure is derived from the weight of successive layers of sediment, while cementation is brought about by the precipitation or growth of minerals (e.g. iron oxide, silica, calcite) around individual particles. Three main classe of sedimentary rocks may be defined according to their transportation history and mode of deposition:

(a) **Clastic rocks**: consist of mineral or rock grains that have been transported as discrete fragments mainly by water (assisted by wind, ice or gravity).
(b) **Precipitated rocks**: are rocks formed as a result of the chemical or organic precipitation of minerals from transported solutions.
(c) **Residual deposits**: are formed in-situ by the weathering or growth of non-transported mateial.

Cross-classification into sub-groups is based on grain-size and/or mineral composition as shown in *Table 3.5*.

Sedimentary rocks are described and identified by a number of characteristics:

(a) *mineral composition:* quartz predominates in sandstones, calcite in limestones, etc.; the nature of the cementing agent is also significant, e.g. an argillaceous sandstone (clay cement) is softer than a siliceous sandstone (quartz cement).
(b) *Grain size and shape:* In BS 5930, the grain-size range of sands (soil) and

TABLE 3.5 Sedimentary rocks

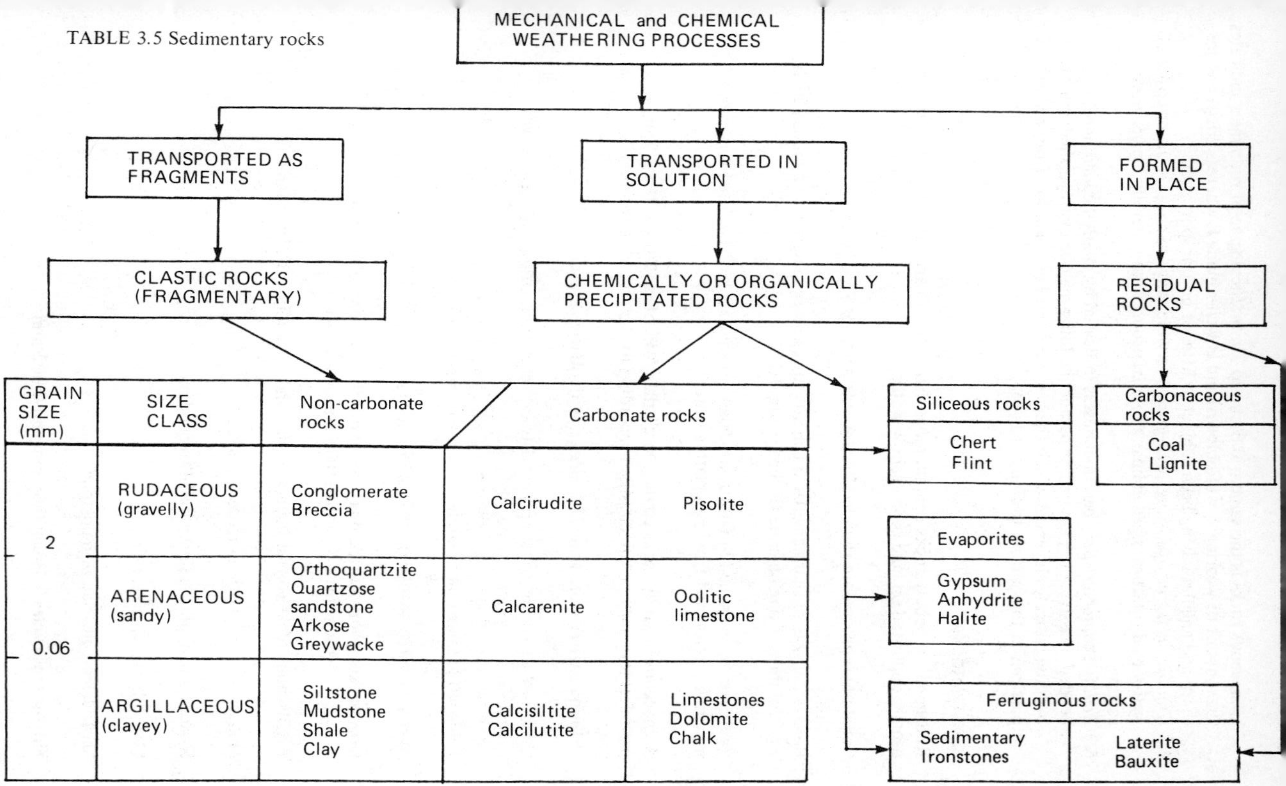

GRAIN SIZE (mm)	SIZE CLASS	Non-carbonate rocks	Carbonate rocks	
	RUDACEOUS (gravelly)	Conglomerate Breccia	Calcirudite	Pisolite
2	ARENACEOUS (sandy)	Orthoquartzite Quartzose sandstone Arkose Greywacke	Calcarenite	Oolitic limestone
0.06	ARGILLACEOUS (clayey)	Siltstone Mudstone Shale Clay	Calcisiltite Calcilutite	Limestones Dolomite Chalk

arenaceous rocks is the same – 0.6 mm to 2 mm; the degree of roundness indicates the amount of wearing – wind-blown sand is well-rounded, whereas **gritstones** are sharp and angular having been subject to little or no transportation.

(c) *Grading or degree of sorting:* **well-graded** (a wide range of sizes) is the opposite of **well-sorted** (uniform-sized grains); more transportation generally means more sorting.

(d) *Bedding characteristics:* bed thickness is indicated by **thinly-bedded, medium-bedded, massive**, etc.; tapered or parallel bedding may be distinguished; graded bedding indicates layers in which there is a continuous gradation from coarse particles at the bottom to fine at the top.

CLASTIC ROCKS

Rudaceous or pebbly rocks contain more than 50% of particles over 2 mm; they are usually deposited on land or in shallow water.

Conglomerates contain rounded (well-worn) particles.

Breccia consists of sharp angular fragments (often scree material) in a muddy matrix, e.g. Dolomitic Conglomerate of the Mendip Hills.

Fault breccia is formed in fault zones when crushed rock is cemented together by minerals precipitated from percolating water.

Arenaceous or sandy rocks contain more than 50% of sand-sized (0.6 – 2mm) particles; they may be sub-classified according to their quartz content:

Orthoquartzite: more than 98% quartz, with quartz cement.

Quartzose sandstone (or simply **sandstone**): predominantly quartz grains, with cement of iron oxide, calcite, clay, etc.

Feldspathic sandstone: mainly quartz, with up to 25% feldspar.

Arkose: quartz, with more than 25% feldspar.

Greywacke: poorly-sorted mixtures of quartz, mica and other minerals; usually $>25\%$ fine-grained (i.e. less than 0.06 mm) material.

Argillaceous rocks consist mainly of clay and mica minerals (less than 0.06 mm).

Mudstone: non-plastic homogeneous rock.

Shale: fissile (splits easily) parallel to bedding.

Clay: plastic when wet.

Marl: calcareous clay or mudstone.

Siltstone: particles between 0.06 mm and 0.002 mm.

CARBONATE ROCKS

Limestone is the term used generally to describe carbonate rocks, the main constituent being calcium carbonate, with other carbonates, notably dolomite, and iron oxide also often present. Many limestones contain clastic, precipitated and organic material (including fossils).

Organic limestones are predominantly of animal or plant origin, such as **shelly-, coral-, reef-** and **crinoidal-limestones. Chalk** consists of the remains of minute marine life (foraminifera and planktonic algae).

Dolomite consists almost entirely of the double carbonate $MgCa(CO_3)_2$, while **magnesian limestone** is mainly calcite with some dolomite or $MgCO_3$.

Oolitic limestone consists of grains of calcite formed by the deposition of several layers around a nucleus of a quartz or shell fragment; the texture resembles fish eggs, hence the term **oolite** or **oolitic.**

Reworked limestone material may be deposited in a similar manner to clastic rocks, producing fragmentary limestones: **calcirudites, calcarenites, calcisiltites** and **calcilutites.**

SILICEOUS ROCKS

These consist mainly of cryptocrystalline silica, such as **chert**, which is found in some limestones, and the irregular nodules of **flint** found in the chalk of Southern England.

FERRUGINOUS ROCKS

These consist of the oxides, carbonates, sulphides or silicates of iron – collectively referred to as **ironstones.** Some are primary deposits, while others are replacements.

EVAPORITES

Evaporites are rock mineral deposits resulting from the evaporation of saline water, such as **gypsum, anhydrite**, and **halite** (See Section 3.3).

RESIDUAL DEPOSITS

Residual deposits have been formed in place without transportation being involved; for example, **carbonaceous** material such as **coal** and **lignite**, and **ferruginous** deposits such as **laterite** and **bauxite.**

3.7 METAMORPHIC ROCKS

Metamorphic rocks are those which have been altered after depositon. The processes of metamorphism are controlled by a number of factors:

(a) temperature changes – both amount and rate of increase or decrease.
(b) pressure changes – relative changes in normal (overburden) and shear stresses.
(c) presence of new fluids or gases.

TABLE 3.6 Metamorphic rocks

Type of metamorphism	Conditions and processes	Examples: Rock	Examples: Texture	Examples: Other features	Grade
Dynamic	Intensely stressed fault zones	FAULT BRECCIA	Cataclastic	Coarse-grained	LOW
	Crushing and shearing	MYLONITE	Cataclastic	Fine-grained, individual crystals fractured.	
Regional	Orogenic (mountain-building) and intrusive uplift areas.	SLATE	Slaty	Highly cleaved product of clay or shale; fine-grained	
	Combined pressure and temperature increase	PHYLLITE	Phyllitic	Medium to coarse-grained; less perfect cleavage than slate.	
	Reorientation, separation and recrystallisation	SCHIST	Schistose	Medium-grained; characteristic foliation & separation of minerals (mica, hornblende, chlorite etc)	MEDIUM
		GNEISS	Banded	Coarsely crystalline product of sedimentary & basic igneous rocks	HIGH
		GRANULITE	Granular	Coarsely-crystalline product of acid-igneous rocks	
Thermal	Adjacent to igneous intrusions	QUARTZITE	Granular	Recrystallised sandstone	
	Temperature increase Recrystallised	MARBLE	Granular	Recrystallised limestone, often 'patterned' due to impurities	
		HORNFELS	Granular	Recrystallised clays and shales	
		SPOTTED SLATE	Slaty	Thermally-altered slate or phyllite	LOW

(d) nature of the existing rock.
(e) time.

There are three main types of metamorphism: **thermal, dynamic** and **regional**; which result in a number of different **textural** types of rock. A general classification of the metamorphic rocks is given in *Table 3.6.*

THERMAL (OR CONTACT) METAMORPHISM

The intrusion of a hot body of magma will often raise the temperature of an existing rock mass sufficiently for it to be **baked** or even completely **recrystallised.** Although there may be an accompanying increase in normal (overburden) pressure, there is usually little or no increase in shear stress.

Alongside large intrusive masses (e.g. granite, gabbro), the alteration zone, called the **metamorphic aureole**, extends outwards, with diminishing effect, up to several kilometres. Common thermal products are **quartzite** from sandstone, **marble** from limestone and **hornfels** from clay or shale. The permeation of magmatic fluids and gases may introduce new elements and compounds, leading to the formation of new (secondary) minerals in the original rocks, e.g. tourmaline, topaz, fluorite.

DYNAMIC METAMORPHISM

In zones of intense pressure, often localised around thrust faults, existing rocks may be crushed and sheared to a high degree. Essentially, dynamic metamorphism is a process of mechanical disruption, with little or no temperature effect.

Common examples are **fault breccias** which are coarse-grained and **mylonite** which consists of very small fractured crystals bound in a rock-flour matrix.

REGIONAL METAMORPHISM

Regional metamorphic processes occur over large areas and are associated with strong crustal movements, such as those producing mountain building adjacent to destructive plate margins. Various **grades** of metamorphism are exhibited in the resultant altered rocks:

Low-grade: occurring near the surface where shear stresses are high, while normal pressure and temperature remain low; exhibit a highly-cleaved **slaty** texture, e.g. **slates, phyllites.**

Medium-grade: occurring at moderate depths with lower shear stress, but higher normal pressure and temperature; exhibit a **foliated** texture, with separation of minerals such as mica, hornblende and chlorite; e.g. **schists.**

High-grade: occurring in deep regions with low shear stress, but with high normal pressure and temperature; high degree of recrystallisation with differential setting giving rise to **banded** textures: e.g.**gneisses.**

METAMORPHIC TEXTURES

Metamorphic rocks are named and described in terms of both the type of metamorphism and their resultant **texture.**

Granular texture:	rounded or sub-angular crystals or grains.
Cataclastic texture:	broken or flaky fragments or crystals.
Slaty texture:	highly-developed parallel cleavage.
Phyllitic texture:	moderate cleavage with some mineral separation and foliation.
Schistose texture:	highly foliated and often contorted, with distinctive mineral separation.
Banded texture:	(sometimes '*gneissic*') mineral separation into lighter and darker bands.

TEST EXERCISES

COMPLETION QUESTIONS (answers on page 205)

Complete the following statements by inserting an appropriate word or words in the spaces indicated.

1 The most abundant elements in the Earth's crust are . . . . and . . . .

2 Na^+, Mg^{2+} and Ca^{2+} are . . . . ions.

3 Cl^-, O^{2-} and S^{2-} are . . . . ions.

4 Two examples of elements which form covalent framework minerals are . . . . and . . . .

5 The majority of rock-forming minerals are compounds of silicon and are collectively called . . . .

6 The way a mineral reflects light from its surface is called its . . . .

7 The way some minerals split along flat smooth surfaces is called . . . .

8 Moh's scale is used to indicate the . . . . of a mineral.

9 The softest mineral is . . . . and the hardest is . . . .

10 Quartz has a hardness of . . . .

11 The most common mineral that cleaves readily into thin plates is either . . . . or . . . .

12 Calcite and galena are minerals that have . . . . cleavage in . . . . directions.

13 Hematite is . . . . oxide and galena is . . . . sulphide.

14 The principal mineral in limestone is . . . .

15 The principal mineral in sandstone is . . . .

16 Granite has an . . . . composition and a . . . . grained texture.

17 Basalt has a . . . . or . . . . grained texture and a . . . . composition.

18 The essential minerals in granite are . . . . , ... and . . . .

19 Kaolin, illite and montmorillonite are known as . . . . minerals.

20 The principal mineral in ultra-basic rocks is . . . .

21 Hornblende and the asbestos minerals, tremolite and crocodilite, are members of the . . . . family.

22 Halite, anhydrite and gypsum are known as . . . .

23 A rock such as Shap Granite which has large crystals set in a fine groundmass is said to have a . . . . texture.

24 Tuffs and agglomerates are known as . . . . deposits.

25 Sedimentary rocks are formed by a dual process of . . . . and . . . .

26 Clastic sedimentary rocks are made up of . . . . cemented together.

27 Precipitated sedimentary rocks are formed by precipitation from . . . .

28 Conglomerates, sandstones and shales are examples of . . . . rocks.

29 Limestones, ironstones and evaporites are examples of . . . . rocks.

30 Shelly, coral, algal, oolitic are all types of . . . .

31 The three types of metamorphism are . . . . , ... and . . . .

32 The zone of altered rock around an igneous intrusion is known as the metamorphic . . . .

33 Some common thermal metamorphic rocks are: . . . . from sandstone and . . . . from limestone.

34 Mylonite and fault-breccias are products of . . . . metamorphism.

35 Highly developed parallel splitting planes are the basis of . . . . cleavage, commonly seen in . . . .

36 A highly foliated and often contorted texture is called . . . .

37 High-grade metamorphic rocks such as . . . . often exhibit a banded texture.

ESSAY QUESTIONS

1 Explain the difference between metallic, non-metallic and inert elements.

2 Explain how mineral compounds are formed ionically and covalently, giving common examples.

3 The most common rock-forming minerals are silicates. State why this should be so and describe the structure of the various types of silicate, giving appropriate examples.

4 State **five** of the physical characteristics by which rock-forming minerals may be identified and give two examples in each case.

5 Discuss the composition, structure and possible origins of the clay minerals.

6 Why are **ferromognesian** minerals so named and in which types of rock would you expect to find them?

7 A large number of metallic ores come from oxides or sulphides. Give a list of the more common of these and discuss their modes of occurrence.

8 Give an account of the origin, composition and texture of a porphyritic granite.

9 Describe the origin and composition of basic and ultrabasic rocks.

10 Discuss the characteristics used to identify and classify sedimentary rocks.

11 Explain the terminology used to describe the various members of the sandstone family: orthoquartzite, quartzose sandstone, feldspathic sandstone, arkose and greywacke.

12 Discuss the origin and composition of limestone and dolomite, explaining the formation of different textures and types.

13 Describe the causes and processes of metamorphism, giving examples of the rocks and minerals occurring in altered rocks.

14 What is meant by **high-grade** and **low-grade** metamorphism? Give examples.

4 Geological structures and mapping

4.1 EARTH-BUILDING PROCESSES

The formations and structures taken up by crustal rocks are the results of the combined effects of many different processes and forces. Although these building-up processes are interactive, it is convenient to group them under three main headings:

(a) *Depositional processes*: such as the laying down of sequences of **strata** in sedimentation, the intrusion of sheets and other igneous masses; giving rise to structural features such as **bedding planes, joints** and **unconformities.**

(b) *Tectonic processes*: processes due to crustal movements; giving rise to uplift and submergence, tilting, folding, faulting, fracturing, vulcanism, etc.

(c) *Erosional processes*: processes due to the weathering and denuding of landsurfaces; giving rise to wearing, cutting, moulding and in other ways forming surface relief.

4.2 STRATIFIED ROCK STRUCTURES

The most notable feature of sedimentary rocks is that they have been laid down in layers or **strata. Stratigraphy** is the study of the sequence or **succession** of stratified rocks in terms of their geological age. In general, sedimentary rock strata consist of parallel **beds**, which are separated by **bedding planes** (*Fig 4.1*). Where a bedding plane separates two different types of rock, it is termed a **strata boundary. Joints** are regular discontinuities occurring at right angles to the bedding planes and usually

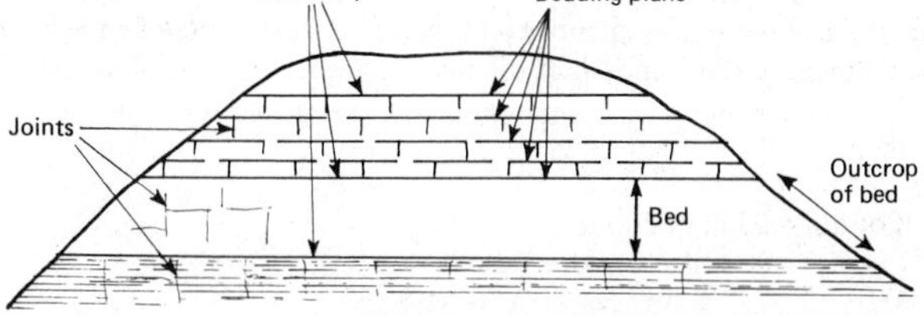

Fig 4.1 Rock strata terms

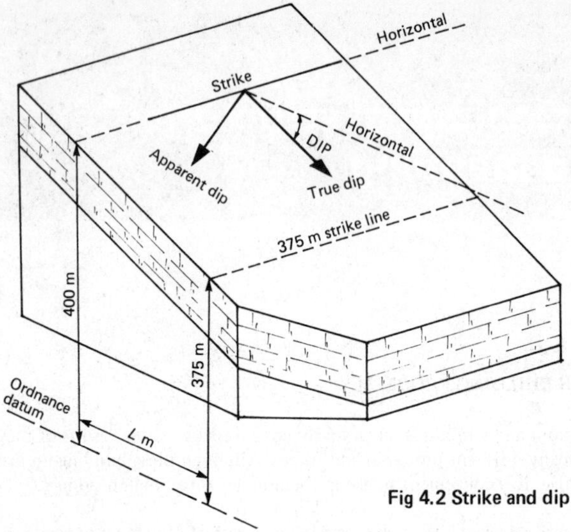

Fig 4.2 Strike and dip

running in two directions at right angles to each other. Rock masses tend therefore (in most cases) to break up naturally into rectangularly-faced blocks, bounded by bedding planes and joints. **Fracturing** is required to bring about any further reduction in size.

An **outcrop** is the area at the ground surface where a rock bed occurs; it is the area shown on a geological map. An **exposure** is any part of an outcrop which may be seen in-situ.

The slope of a tilted bed of rock is defined by its **strike** and **dip** (*Fig 4.2*).

Strike: is the direction in which horizontal lines could be drawn on the bedding plane.

Strike lines: may be drawn at prescribed heights above ordnance datum and so represent the bedding plane in the same way that surface contours represent the ground surface.

Dip: is the maximum angle below horizontal made by the bedding plane; the direction of dip is therefore perpendicular to the strike.

Apparent dip: is the dip of the bedding plane measured in a given direction, other than perpendicular to the strike.

The dip of a bed is a vector quantity and therefore must be stated in terms of both **amount** and **direction.** The amount of dip may be given either as a slope ratio or as an angle: for example, in *Fig 4.2*, suppose the interval (L) between the two strike lines is 200 m, then:

$$\text{Amount of dip} = 25 \text{ m in } 200 \text{ m}$$
$$= 1:8 \quad \text{(slope ratio)}$$
$$= \arctan \tfrac{1}{8} = 7° \text{ (angle)}$$

The direction of dip is given as a bearing from true north, and will be at right-angles to the strike. On a geological map a given structure will produce a characteristic pattern of outcrops.

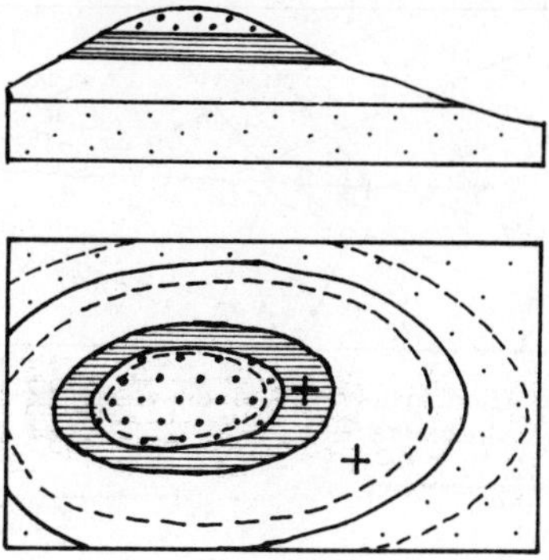

(a) Horizontal strata

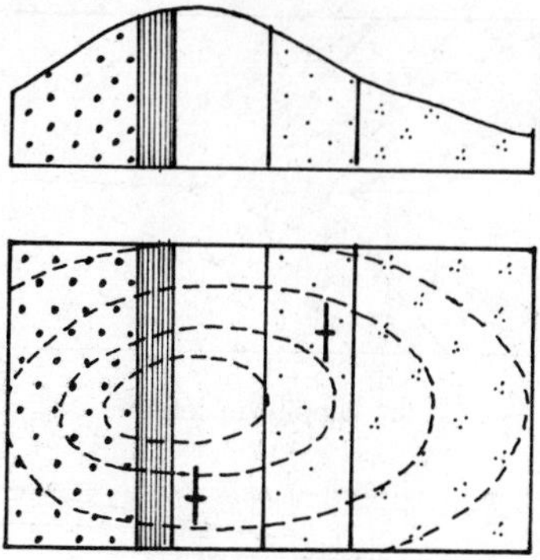

(b) Vertical strata

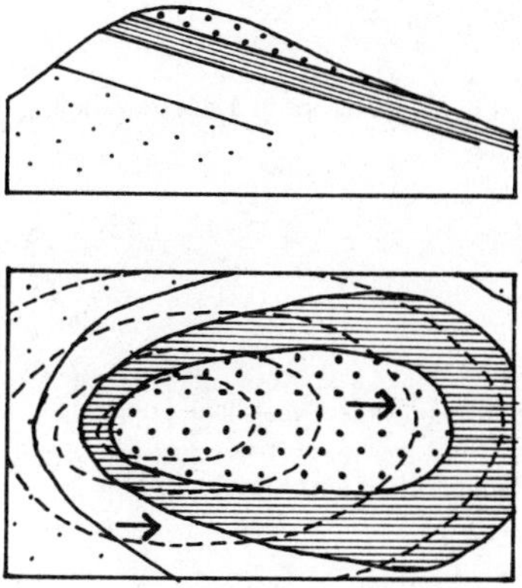

(c) Constant dip and strike

Fig 4.3 Outcrop patterns

Horizontal strata: the outcrops will run parallel to the surface contours; indicated by a + sign (*Fig 4.3a*).

Vertical strata: the outcrops will be straight lines and parallel to each other; indicated by —|— sign. (The longer line showing the direction of strike) (*Fig 4.3b*).

Dipping strata: strikes lines must be drawn to determine the direction and amount of dip (*Fig 4.3c*).

It should be remembered that since outcrops occur at the ground surface they reflect the shape of the ground. For example, on a gently rounded hill, the outcrop has a

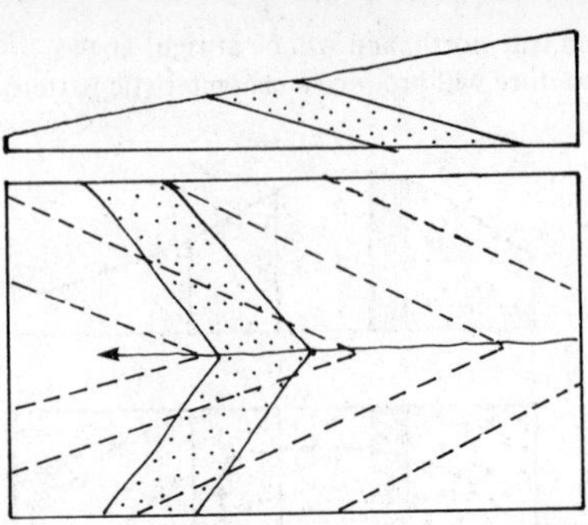

(a) Dipping upstream

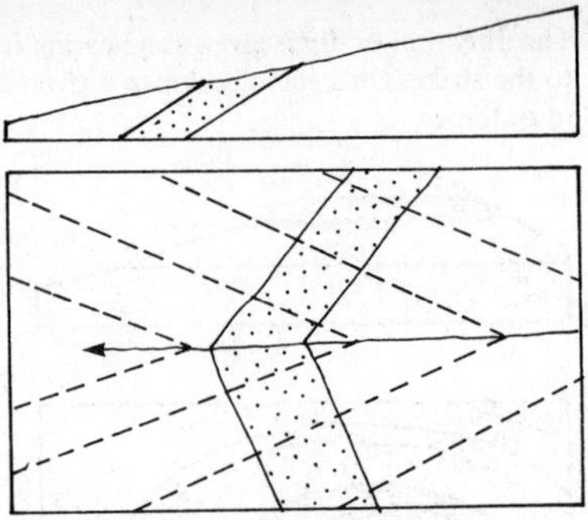

(b) Dipping downstream: dip greater than stream gradient

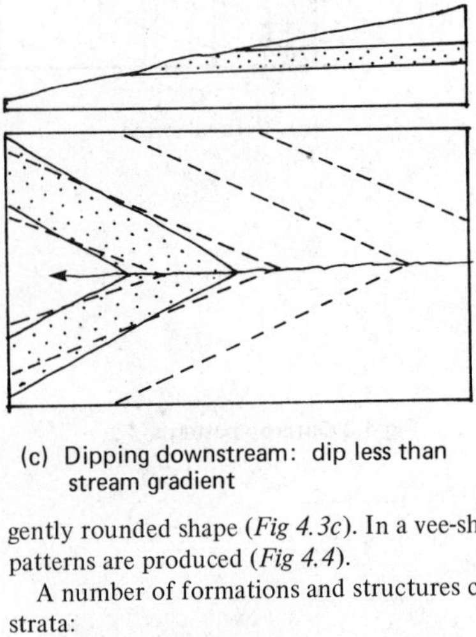

(c) Dipping downstream: dip less than stream gradient

Fig 4.4 Outcrops in a stream valley

gently rounded shape (*Fig 4.3c*). In a vee-shaped stream valley, vee-shaped outcrop patterns are produced (*Fig 4.4*).

A number of formations and structures can be identified with simply dipping strata:

Dip slope: ground surface sloping parallel with the dip of the underlying beds (*Fig 4.5*).

Scarp (or escarpment): hillside slope cutting across the bedding planes (*Fig 4.5*).

Cuesta: a dip-slope and scarp hill or ridge (*Fig 4.5*).

Mesa: flat-topped hill with scarps on either side (*Fig 4.6*).

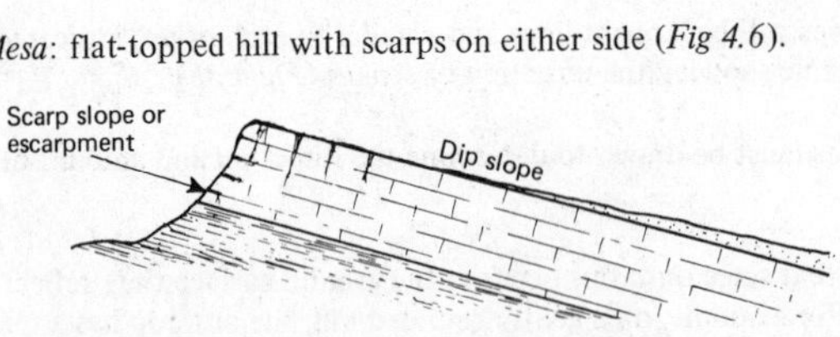

Fig 4.5 A cuesta

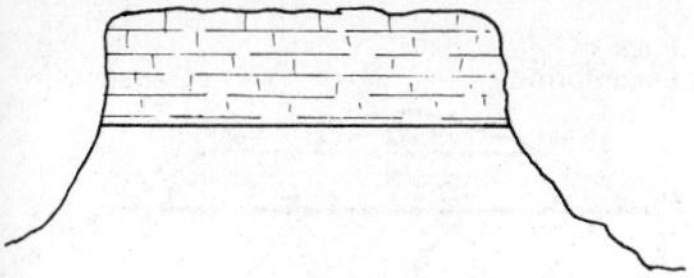

Fig 4.6 A mesa

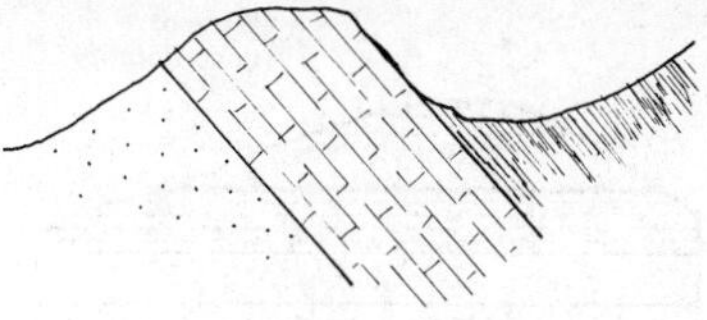

Fig 4.7 A hog's back

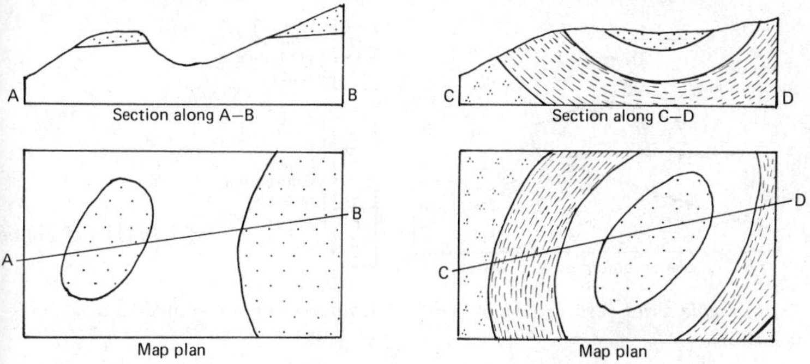

(a) Formed by erosion

Fig 4.8 Outliers

(b) Formed by folding

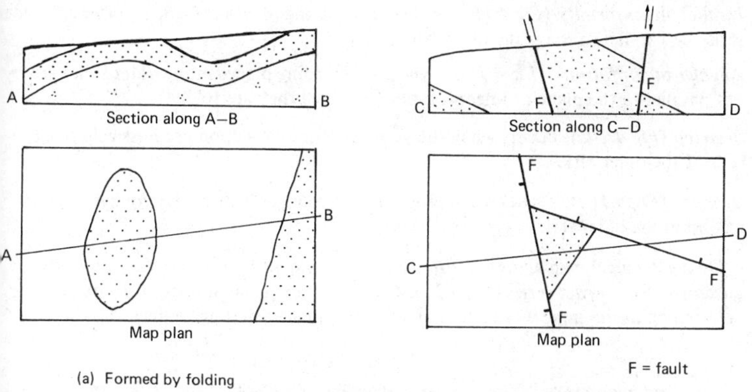

(a) Formed by folding

Fig 4.9 Inliers

(b) Formed by faulting

Hog's back: hill or ridge with dip-slope and scarp each at 45° (*Fig 4.7*).

Outlier: an outcrop of young rock surrounded by older rock; formed either by erosion, folding or faulting (*Fig 4.8*).

Inlier: an outcrop of old rock surrounded by younger rocks; formed either by erosion, folding or faulting (*Fig 4.9*).

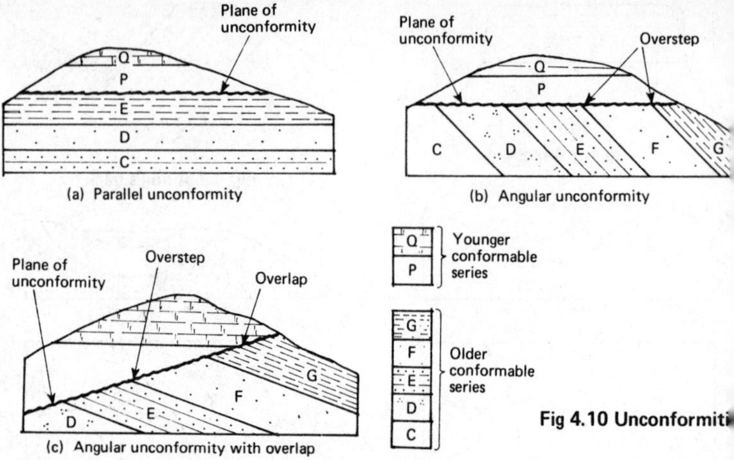

Fig 4.10 Unconformiti

Conformable series: a succession of beds laid down in an **uninterrupted** sequence of deposition.

Unconformity: a break in the depositional sequence, indicating a period of erosion; se as a plane (or reasonably regular) surface of separation between older and younger series; usually represents an ancient erosion surface (*Fig 4.10*).

Parallel unconformity (*Fig 4.10a*): where the bedding planes in both younger and old series lie parallel to the plane of unconformity.

Angular unconformity (*Fig 4.10b*): where the bedding planes in the older series are not parallel to the plane of unconformity (or where they are folded).

Overstep (*Fig 4.10c*): occurs when the younger series rests upon progressively older beds of the older series.

Overlap: (*Fig 4.10c*): occurs when progressively younger beds of the younger series rest upon the older series.

On a geological map an unconformity is made evident by the interruption of the outcrops of the lower series (*Fig 4.11*). The dip of the plane of unconformity is determined in the same way as that of a bedding plane using strike lines.

Worked example 4.1 The geological map given in *Fig 4.11* shows an area in which a bed of limestone (bed Y) lies unconformably on a series of older beds (beds P, Q, R, S, T, U).

(a) Indicate on the map the outcrop of the unconformity.
(b) Assuming it to have constant dip and strike, determine the amount and direction of dip of the plane of unconformity.
(c) Assuming constant dip and strike, determine the amount and direction of dip of the boundary plane between beds P and Q.
(d) Determine the thickness of each of the beds in the lower (older) series.

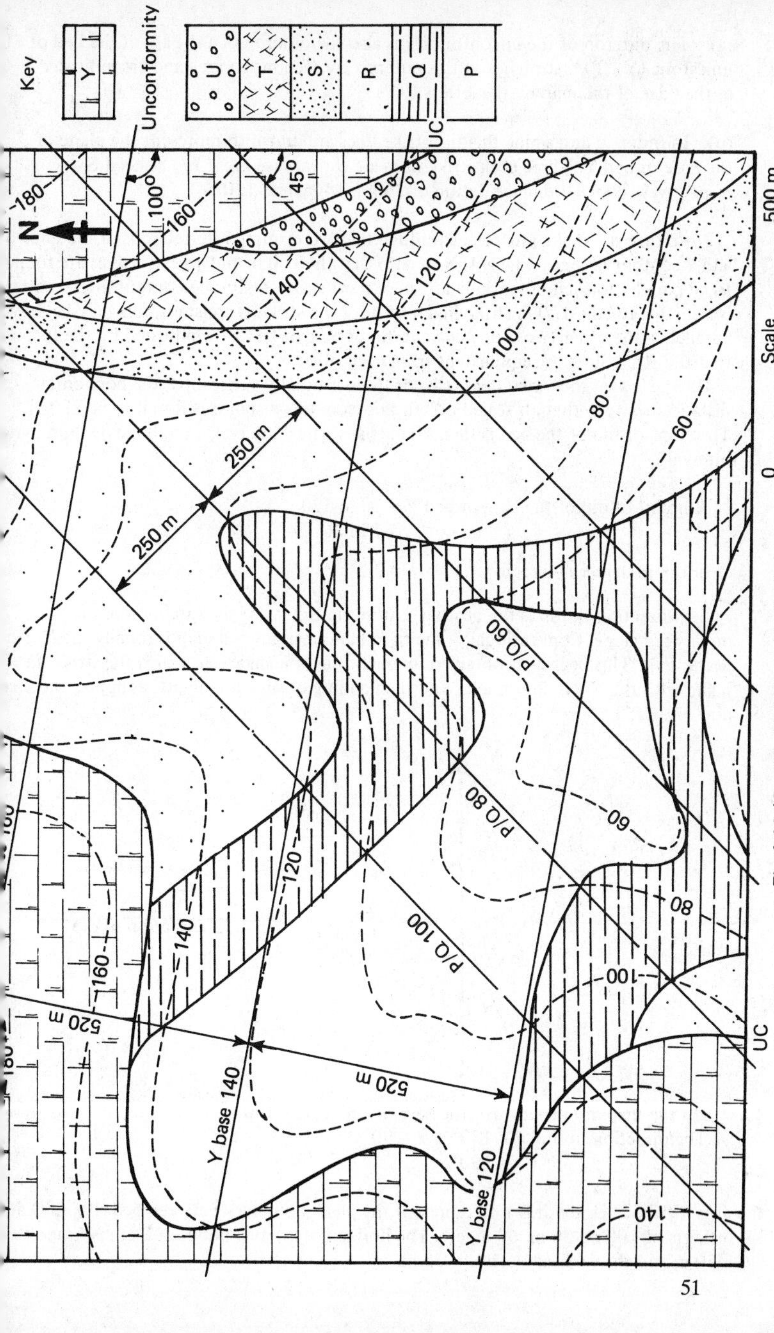

Fig 4.11 (See Worked Example 4.1)

(a) The outcrop of the unconformity is also the outcrop of the base of the bed of limestone (Y). Two stretches of this outcrop are seen in the area and are indicated at the edge of the map by the letters UC.

(b) In order to determine the dip, strike lines are drawn to represent the plane of the base of bed Y. Three such strike lines are shown (see *Fig 4.11*) at heights respectively above ordnance datum of 160m, 140m and 120m.

Note that the strike line passes through the two or more intersections of the outcro and a contour of a given height. Note also that the 160 m and 140 m strikes are found to be parallel to each other, indicating constant strike. Only one contour intersection is available for the 120 m strike line, but since the strike is constant, this may be draw parallel to the other two. The constant dip of base of bed Y is indicated by the fact that the strike lines are equidistant from each other.

Having drawn the strike lines (two or three is usually sufficient), the horizontal distance between them is scaled off. In this case, this distance is found to be 520 m. The slope or dip of the bed is then 20 m (the vertical interval) in 520 m (the horizont interval).

Amount of dip of the base of bed Y = 20 in 520
= **1 in 26**

or dip in degrees = arctan $\frac{1}{26}$ = 2.2°

The **direction** of dip is the bearing (from true North) of the direction of maximum downward slope. Conventionally, the right-hand margin on a map is usually the North South axis. This bearing is obtained by measuring the angle (α), which the strike lines make with the North-South axis, and then adding or subtracting 90° as appropriate (se also *Fig 4.12*).

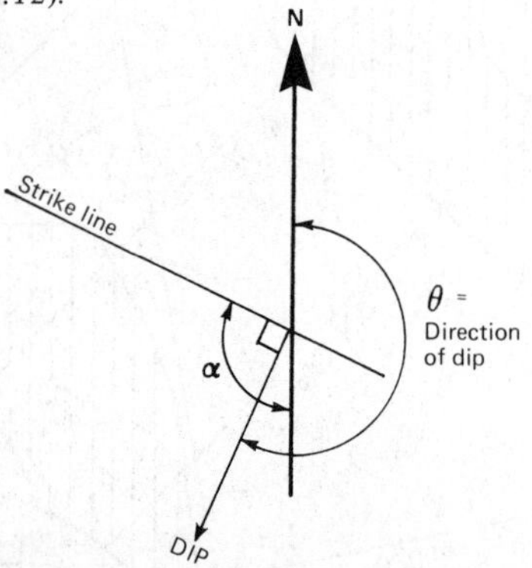

Fig 4.12 Direction of dip

On the map in *Fig 4.11*, for the base of bed Y, $\alpha = 100°$
The direction of dip, $\theta = 180° + \alpha - 90°$
= **190°**

(c) Strike lines are drawn to represent the plane boundary between beds P and Q at heights of 100 m, 80 m, 60 m, etc. The horizontal interval between these is found to 250 m and the vertical interval is 20 m.

The dip of boundary P/Q = 20 in 250
= **1 in 12.5**

or dip in degrees $= \arctan \dfrac{1}{12.5} = 4.6^\circ$

The direction of this dip $= 180^\circ + 45^\circ - 90^\circ$
$= \mathbf{135^\circ}$

(d) The vertical thickness of a bed may be determined by comparing strike lines drawn on its upper surface with those drawn on its lower surface.

In *Fig 4.11* the lines drawn to represent the 60 m, 80 m, and 100 m strike lines for surface P/Q will be seen to also represent respectively the 80 m, 100 m and 120 m strike lines for surface Q/R. To verify this, trace along the outcrop of surface Q/R and note the contour values at the strike line intersections. The strike lines representing Q/R are exactly 10 m above those for P/Q; therefore the thickness of bed Q is 10 m. Similarly the other bed thicknesses are found to be:

Bed R : 20 m
Bed S : 10 m
Bed T : 10 m

Since the complete thickness of beds P and U cannot be seen on this map, it is only possible to give an estimated minimum value:

Bed P : >9 m
Bed U : >8 m

Similarly bed Y is at least 21 m thick.

4.3 FOLDING

Masses of rock which are subject to **compressive forces** may buckle or they may fracture. Fracturing is more likely to occur near the surface and when the forces are applied quickly. Deeper in the crust, where confining pressure is high, and when the forces are applied very slowly over long periods of time, **folds** may be formed due to a process of plastic flow.

Competent strata are more resistant and ductile and therefore bend and fold with little or no internal disruption.

Incompetent strata are weaker and more brittle by comparison and therefore may be disrupted by internal shear stress, e.g. slaty cleavage may develop.

An **anticline** is formed when the strata are folded upwards into an arch shape and a **syncline** when they are folded downward into a trough (*Fig 4.13*). The **axial plane** bisects the angle between the limits of the fold and the **axial trace** is the outcrop (hypothetical) of the axial plane. On a geological map an anticline will be evident from the fact that dip arrows will point outwards (away from the axial plane); in the case of a syncline they point inwards.

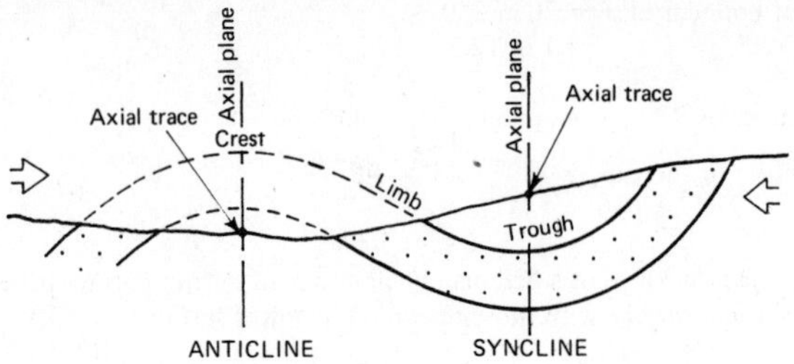

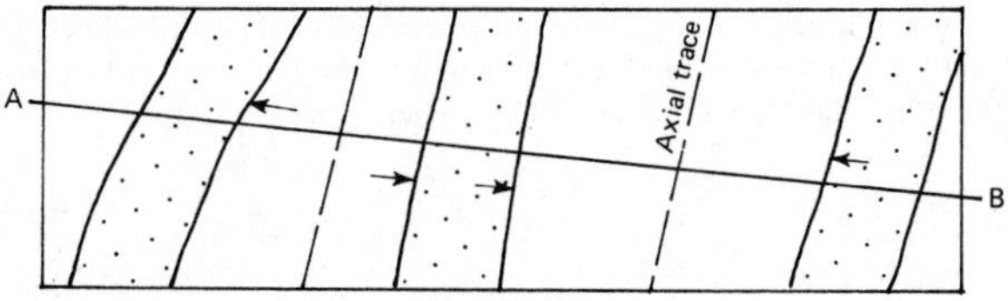

Fig 4.13 Anticline and syncline

Fig 4.14 shows a sequence of fold structures, becoming more complex as the degree of compression increases. Although the diagrams show principally anticlinal forms, similar terminology applies to synclinal folds.

If the axial plane is inclined to horizontal, the fold is said to **plunge** (*Fig 4.15a*). In a **pericline** the axis plunges in opposite directions away from the highest point of the structure (*Fig 4.15b*). In a **dome** the beds tend to dip almost equally away from the centre in all directions (*Fig 4.15c*).

In the mapping of fold structures a set of strike lines for each limb is drawn on either side of the axis (*Fig 4.16*)

Worked example 4.2 Determine the dip of each limb of the fold shown in *Fig 4.16* from the strike lines drawn. State the type of fold.

Scaling off from the map, the strike-line horizontal intervals are:

LH LIMB = 125 m
RH LIMB = 75 m

The strike-line vertical interval is 20 m in both

Hence, dip of LH LIMB = 20 : 125 = 1:6.25 = **9° almost due W**
dip of RH LIMB = 20 : 75 = 1:3.75 = **15° almost due E**

This fold is therefore an **asymmetric anticline**.

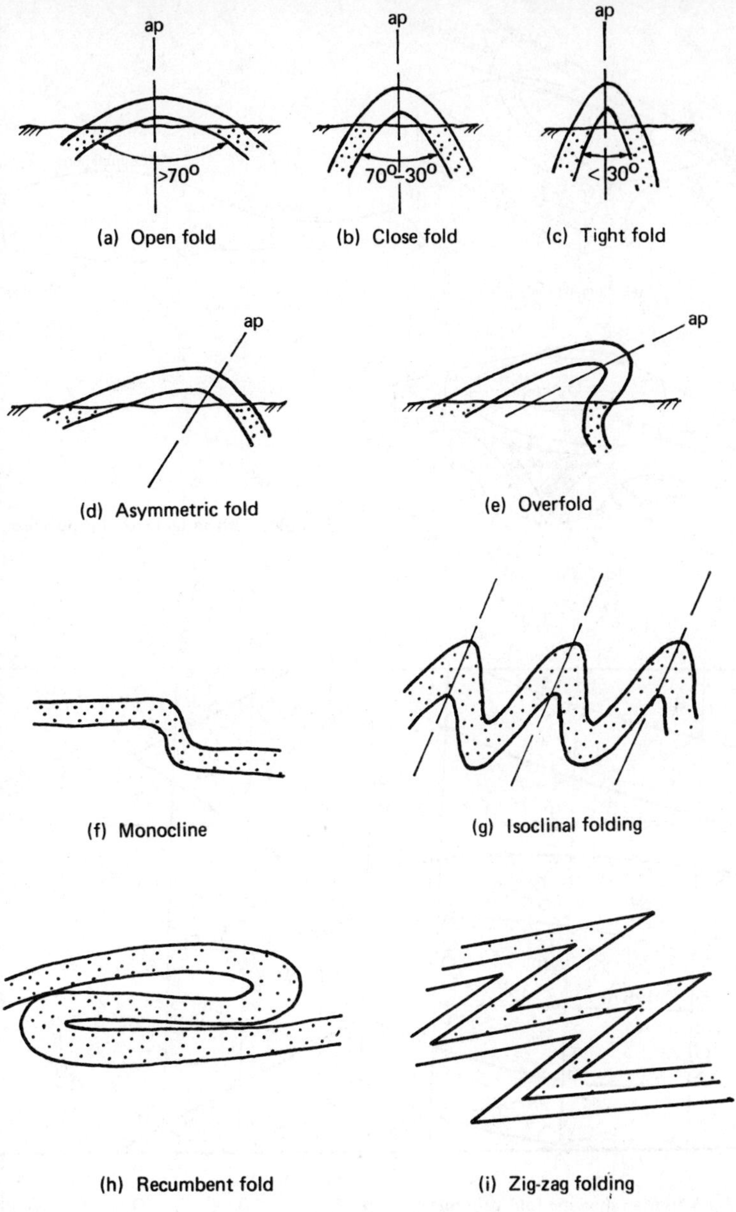

Fig 4.14 Types of fold

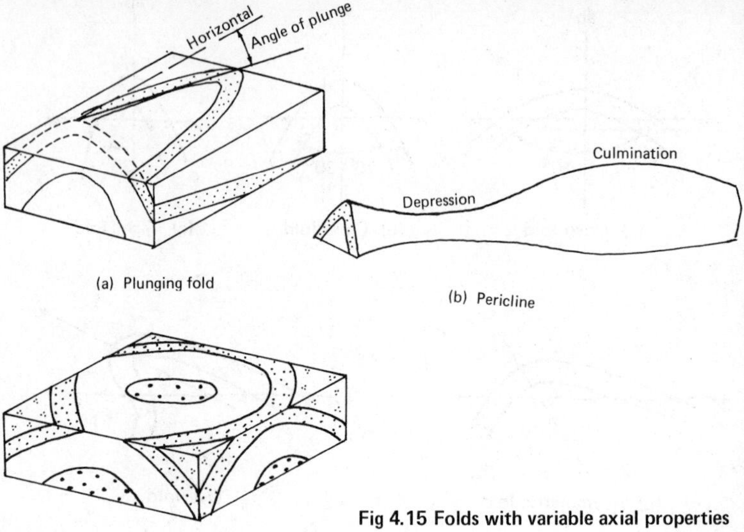

(a) Plunging fold

(b) Pericline

(c) Dome

Fig 4.15 Folds with variable axial properties

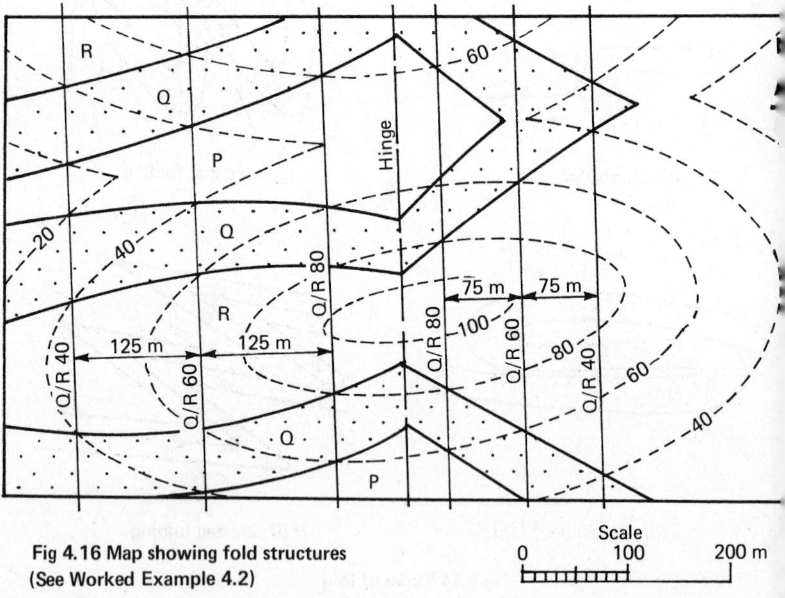

Fig 4.16 Map showing fold structures (See Worked Example 4.2)

4.4 FAULTING

A **fault** is formed when rocks are fractured along a shear plane (**fault plane**), so that the beds on one side are displaced relatively to those on the other. The formational attitude of a fault plane, and the relative movement, is described by the following terms:

Dip and strike: (of the fault plane) have the same meaning as for a bedding plane.

Throw: is the vertical displacement of one side of the fault relative to the other.

Strike-slip: (or wrench component) is the relative horizontal displacement on either side of the fault.

Hade: is the angle made between the fault plane and vertical, i.e. hade = 90°– dip.

Three main types of fault can be defined in terms of both their mode of formation and their subsequent structural characteristics: **Normal faults, reverse faults** and **wrench faults.**

NORMAL FAULTS

The majority of faults (hence the term **normal**) develop as shown in *Fig 4.17a and b*, either when horizontal tension is the dominant force (the more general case), or when the vertical compression is much greater than the horizontal compression.

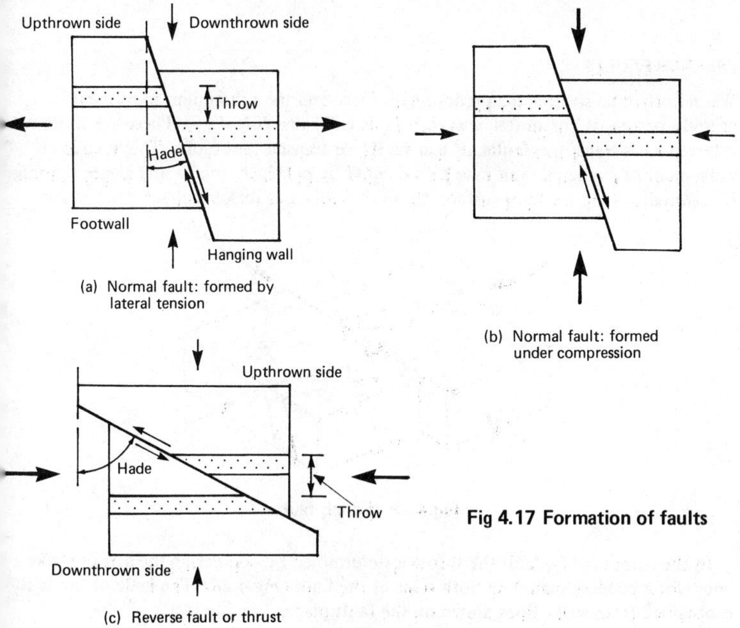

(a) Normal fault: formed by lateral tension

(b) Normal fault: formed under compression

(c) Reverse fault or thrust

Fig 4.17 Formation of faults

The principal structural characteristics are a fault plane which is entirely vertical or with the **hade towards the downthrow side.** It is also apparent that the sheared edges of the beds have moved away from each other on either side of the fault plane.

REVERSE FAULT (OR THRUST)

When horizontal compression is the dominant force a reverse fault is formed (*Fig 4.17c*). These are often referred to as **thrusts** or **thrust faults** or **overthrusts,** especially when the fault plane lies at a flat angle.

The hade of a reverse fault is **towards the upthrow side.** It is also apparent that the sheared edges of the beds have overlapped each other at the fault plane. A thrust fault is often an ultimate development of a recumbent fold (*Fig 4.18*).

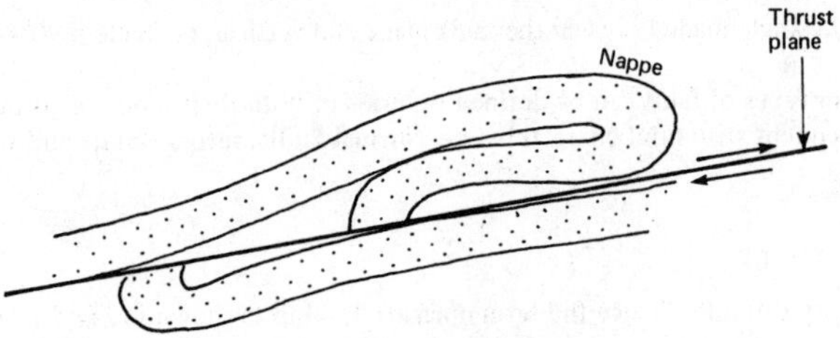

Fig 4.18 Overthrust developed from a recumbent fold

WRENCH FAULTS

When horizontal shear is the predominant force and the subsequent movement is entirely or almost horizontal, a **wrench fault** is formed (*Fig 4.19*). These are alternative referred to as **transform faults,** or **tear faults,** or **transcurrent faults.** The horizontal movement of a wrench fault may be indicated by polished grooves and ridges running horizontally along the fault surface; these are known as **slickenslides.**

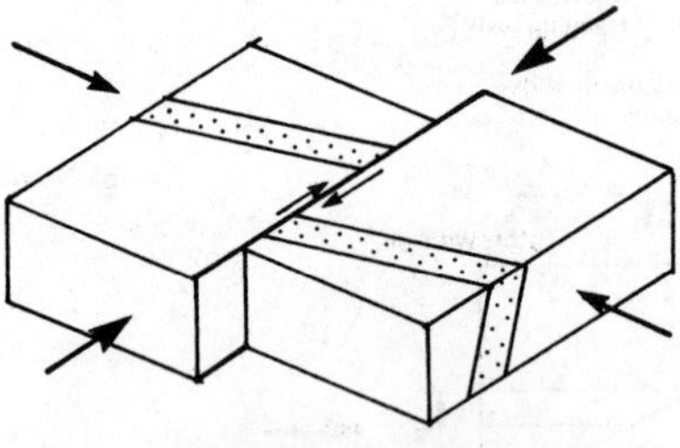

Fig 4.19 Wrench fault

In the mapping of a fault the throw is determined by comparing the sets of strike lines (for a bedding plane) on both sides of the fault (*Fig 4.20*). The hade of the fault is obtained from strike lines drawn on the fault plane.

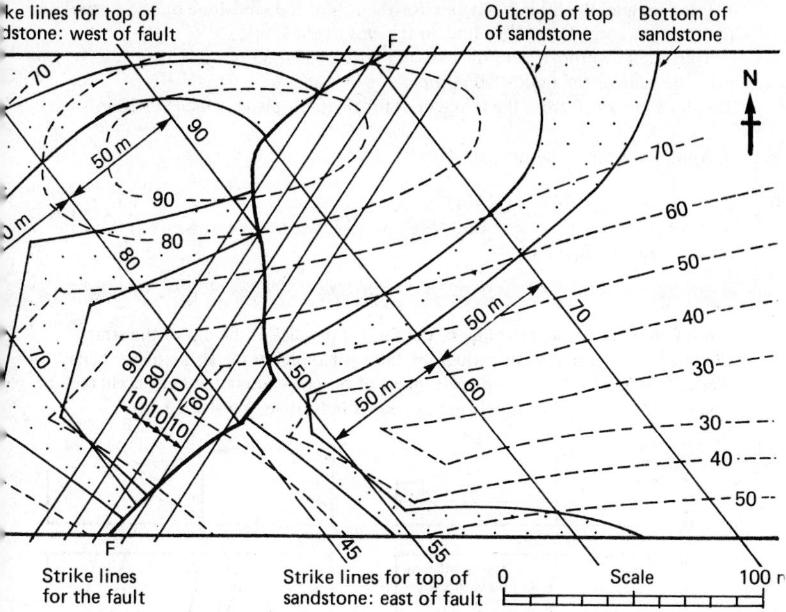

Fig 4.20 Map showing fault structure (See Worked Example 4.3)

Worked example 4.3 In the area shown by the map in *Fig 4.20* a fault cuts through a bed of sandstone. On the assumption that the strike and dip of both the sandstone and the fault are constant, strike lines have been drawn as shown. Determine:

(a) the amount of dip of the sandstone;
(b) the thickness of the sandstone;
(c) the amount and orientation of the downthrow of the fault;
(d) the angle of hade of the fault;
(e) whether it is a normal or a reverse fault.

(a) The strike lines drawn for the top of the sandstone are found to be 50 m apart, the vertical interval being 10 m.

Amount of dip = 10 : 50 = **1 : 5 = 11°**

(b) Examination of the strike lines for the top of the sandstone shows that they also represent the strike lines for the bottom of the sandstone, at heights 10 m lower.

For example, the 70 m strike line can be seen to pass through the 60 m contour and the outcrop of the bottom bedding plane of the sandstone. Thus the vertical thickness of the sandstone = **10 m.**

(c) The downthrow of the fault is obtained by comparing the displacement of a strike line on either side of the fault.

For example, the 80 m strike line for the top of the sandstone on the west of the fault becomes the 45 m strike line on the east of the fault.

Thus the downthrow is to the east and is 80 – 45 = **35 m.**

(d) the hade of the fault = 90° – dip.

The dip is obtained from the strike lines for the fault plane, which are 10 m apart.

Amount of dip of fault = 10 : 10 = 1 : 1
= 45°

Angle of hade = 90° – 45° = **45°**

(e) Since the fault hades towards the downthrown side, i.e. the east in this case, it is therefore a **normal** fault.

A **dip** fault occurs when the line of the fault is parallel to the dip of the strata (*Fig 4.21a*) and a **strike fault** when the fault is parallel to the strike of the beds (*Fig 4.21b*). On the map the displacement of outcrops along the fault is characteristic of a dip fault, while repetition of beds is indicative of a strike fault.

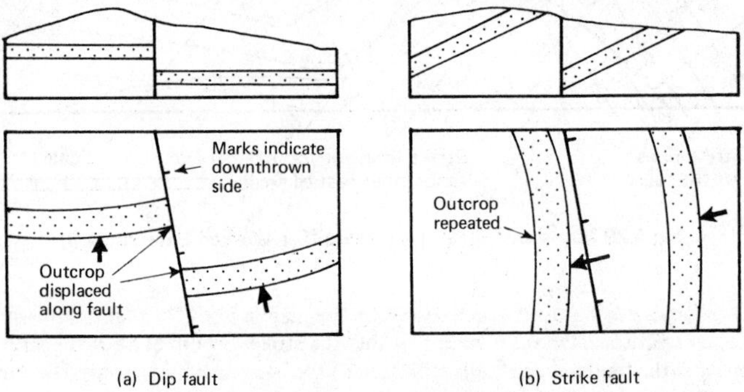

Fig 4.21 Dip and strike faults

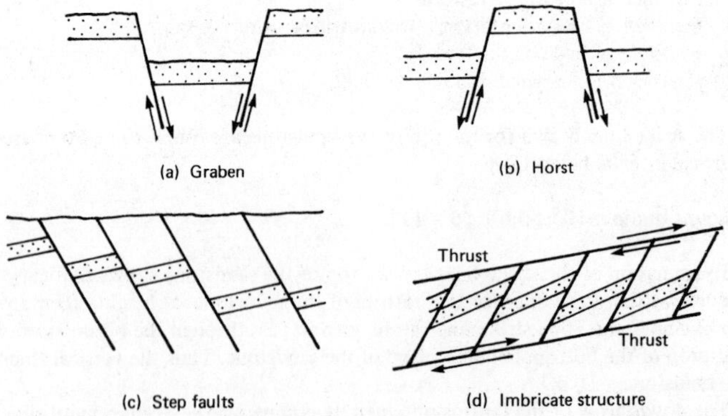

Fig 4.22 Multiple fault structures

FAULT GROUPS

Where several faults occur in close proximity, a group structure is recognisable (*Fig 4.22*).

Graben or rift: a downfaulted block between two normal faults; a **rift valley** is formed following erosion.

Horst: an upthrown ridge between two normal faults.

Step faults: a parallel series of normal faults downthrowing in the same direction.

Imbricate structure: a series of steep reverse faults developed between two parallel flat thrusts.

4.5 IGNEOUS ROCK STRUCTURES

Igneous rock formations will be either **concordant** or **discordant** with the sedimentary structures in an area.

Concordant formations – lie within the sedimentary rock succession, e.g. sills, lava flows, tuff beds.

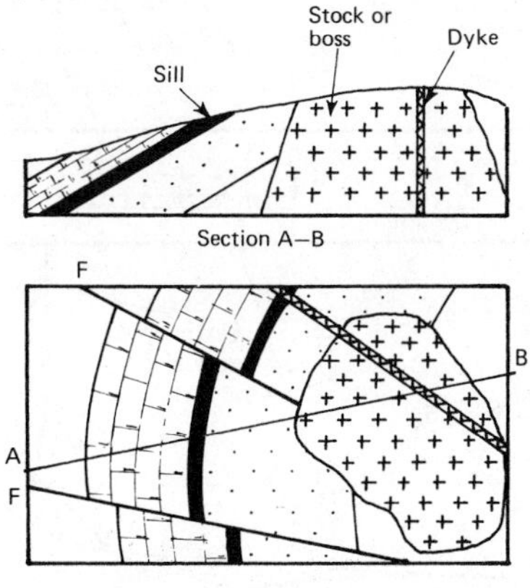

Fig 4.23 Igneous rock outcrops

Discordant formations – cut across the sedimentary rock bedding planes, e.g. dykes, stocks, bosses, necks, batholiths.

The main types of igneous rock outcrops are shown in *Fig 4.23*.

4.6 PLOTTING A GEOLOGICAL MAP

The information from which map plotting data is derived includes field survey details of outcrop exposures, exposures in cuttings and excavations, borehole logs, geophysical data, and so on. The basic procedure is the same in all cases and starts with the drawing of strike lines.

In *Fig 4.24*, suppose a particular bedding plane of constant dip and strike is known to outcrop and points A, B and C. First, a triangle is drawn with A, B and C at corners. Each side is now subdivided so that the intervals correspond with the contour interval (i.e. 25 m, 20 m, 10 m, as the case may be). Now, by joining points of the

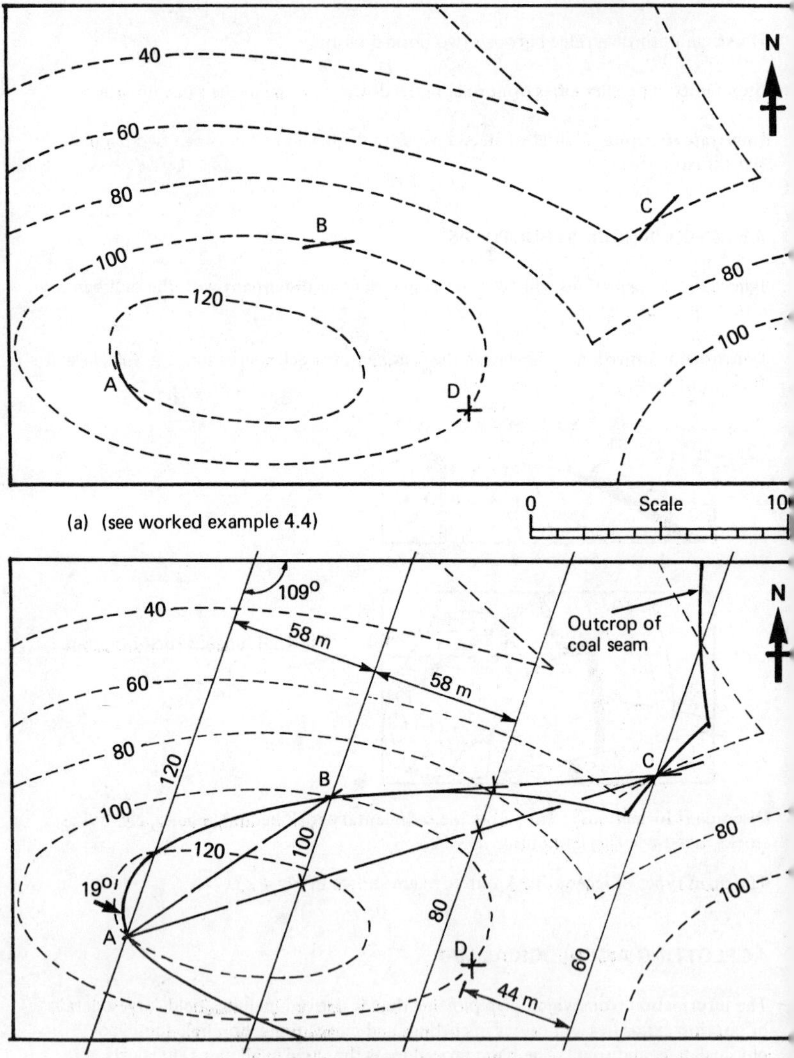

(a) (see worked example 4.4)

(b)

Fig 4.24 Map showing three-point construction

same height on the triangle, some of the strike lines are drawn and others added parallel to these as required. The outcrop is plotted between the intersection of the two sets of lines: contours and strike lines. That is, a point on the outcrop occurs where a contour and a strike of the **same height value** intersect. Additional sets of strike lines can be added for other bedding planes, unconformities, fault planes, etc.

Worked example 4.4. In the area shown in *Fig 4.24a*, the outcrop of a coal seam is seen to occur at points A, B and C. Assuming the coal seam to have constant strike and dip:
(a) determine the amount and direction of its dip from a three-point construction;
(b) draw the strike lines and complete the outcrop of the coal seam in the area;
(c) determine the depth below ground surface of the seam at point D.

(a) The first step in the three-point construction is to draw a triangle with points A, B and C at its corners. Since each corner is at an outcrop, the triangle is drawn on the bedding plane of the coal seam. The height of each corner is deduced from the contours:

A = 120 m + OD; B = 100 m + OD; C = 60 m + OD (*Fig 4.24b*).

Now a point midway between B and C must be 80 m + OD and the two points equally spaced between A and C must be 100 m + OD and 80 m + OD, respectively. Strike lines are now drawn through points of equal height, e.g. at 100 m + OD and 80 m + OD.

The strike-line horizontal interval is found to be 58 m by scaling; the vertical interval is 20 m.

Hence the amount of dip = 20 : 58 = **1 : 2.9 = 19°**

From the orientation of the strike lines the direction of dip = **109°**.

(b) When the 120 m, 100 m, 80 m and 60 m strike lines have been drawn, the outcrop of the coal seam can be completed.

(c) Since the seam dips at a slope of 1 in 2.9, then in 44 m it rises 15 m. Hence at D it is **75 m + OD, i.e. 25 m below the ground surface.**

4.7 PUBLISHED GEOLOGICAL MAPS

Geological maps are produced for a variety of purposes, at a variety of scales and in a variety of forms. Black and white engineering maps are usually incorporated into a geological survey report for a given site or area. The information given may be of a structural nature, or geomorpholoigcal, or economic (relating to construction material or mineral deposits), or hydrogeological. Not all of such maps will be generally published, and many will be restricted to a few interested parties.

The Institute of Geological Sciences produce a wide range of maps for public purchase, the majority of which may be obtained from the Bookstall at the Museum in Kensington, London. The following examples give some indication of the diversity of scales and types:

ONE-INCH (OR 1:50 000) SERIES MAPS

The most widely used of the published maps are those in the One-Inch to One-Mile Series, (now being republished at the SI scale of 1:50 000). There are three sets, one each for England and Wales (together), Scotland and Northern Ireland. It is essential that students be given an opportunity to study some of the One Inch sheets.

In areas where the amount of glacial and other superficial deposits are significantly high, two editions of each sheet are published.

Solid edition: in which the outcrops of the superficial (i.e. **drift**) deposits are left uncoloured, thus showing the underlying **solid** rock surface. (Note: the drift **boundaries** are also printed).

Drift edition: in which the superficial deposit outcrops are characteristically coloured, but with all of the sub-drift solid boundaries also printed.

A single **solid-with-drift** edition is published of sheets where only a small amount of drift occurs. In this edition, the drift outcrops are coloured as for the **drift** edition, but the underlying solid boundaries are not drawn.

1:10 000 SERIES MAPS

The 1:10 000 series of maps (replacing the 6 inch to 1 mile series) are basically ordnance survey maps of a 5 km square area overprinted with geological information. These are available for mining and other areas of special interest.

REGIONAL GEOLOGY HANDBOOKS AND SHEET MEMOIRS

The Institute of Geological Sciences also published a series of **handbooks** in which the geology of a **region** (i.e. South-West England, London and Thames Valley, South Wales, Grampians, Highlands, etc.) is described, together with maps and photographs. For some of the one-inch sheets a **Sheet Memoir** is published describing in detail the geology of the area of a particular sheet. More detail, including borehole records and survey data, is given in a Sheet Memoir. Both of these publications will provide valuable information relating to the geology of a site.

TEST EXERCISES

COMPLETION QUESTIONS (answers on page 205)

Complete the following statements by inserting the most appropriate word or words in the spaces indicated.

1 The three main groups of earth-building processes are . . . . , ... and . . . .

2 Sedimentary rock strata generally consist of parallel . . . . separated by . . . .

3 Discontinuities running at right-angles to each other and roughly perpendicular to the direction of bedding are called . . . .

4 An outcrop occurs where a rock . . . . meets the ground surface.

5 On a bedding plane the direction of horizontal is called the . . . .

6 Lines drawn on a bedding surface which connect points of equal height above ordnance datum are called . . . . lines.

7 Dip is the maximum . . . . below . . . . made by a bedding plane.

8 The dip of a bed must be given in terms of both . . . . and . . . .

9 The following signs are used on a geological map to indicate, respectively, dip, horizontal strata and vertical strata: . . . . , ... and . . . .

10 State the structural terms indicated in *Fig 4.25*
A . . . . , B . . . . , C . . . .

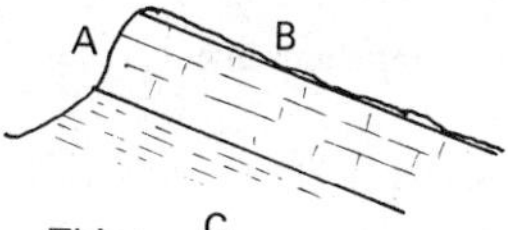

Fig 4.25

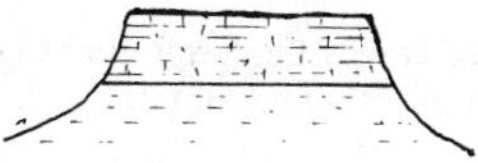

Fig 4.26

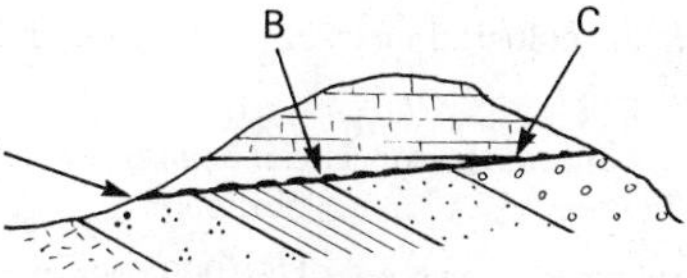

Fig 4.27

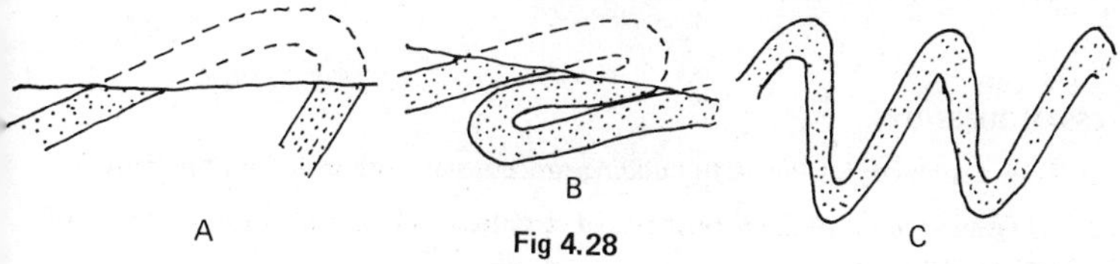

Fig 4.28

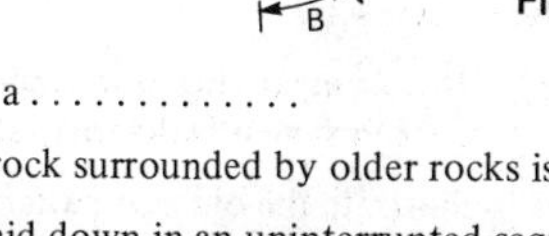

Fig 4.29

11 *Fig 4.26* is a sketch of a . . . .

12 An outcrop of young rock surrounded by older rocks is called an . . . .

13 A succession of beds laid down in an uninterrupted sequence is called a . . . . series.

14 State the structural terms indicated in *Fig 4.27*
A . . . . B . . . . C . . . .

15 .... strata bend or fold easily, whereas .... strata tend to fracture.

16 Arch-shaped folds are called .... and trough-shaped folds ....

17 In a fold, the angle between adjacent limbs is bisected by the ....

18 State the types of fold indicated by the diagrams in *Fig 4.28*
A .... B .... C ....

19 In a pericline or a dome the beds dip .... the centre in all directions.

20 A fault is formed when the rocks .... along a .... plane.

21 The terms describing the relative displacement at a fault indicated in *Fig 4.29* are:
A .... B .... C ....

22 A normal fault hades .... the downthrown side.

23 A thrust or overthrust is another name for a flat .... fault.

24 Give two of the terms used to describe a fault at which the movement is predominantly horizontal: .... , ...

25 On a geological map the same beds may appear repeated on either side of a .... fault.

26 The One-inch series of geological maps of Great Britain is being replaced by the 1 in .... series.

27 In a .... edition, the superficial deposits are left uncoloured, whereas they are coloured in the .... edition.

28 A line of 0.2 mm thickness drawn on a 1/50 000 scale map represents a thickness of .... metres on the ground.

ESSAY QUESTIONS

1 Give a summary of the earth-building processes at work in and on the crust.

2 Give an account of the formations of stratified rocks in which various forms of unconformity occur.

3 Show by means of annotated sketches the principal features in dip-slope and scarp scenery.

4 Explain, using appropriate diagrams, the sequence of folding and faulting that migh give rise to a recumbent fold with overthrust.

5 The San Andreas fault is an active transcurrent fault. Discuss this statement from the structural geology point of view.

6 Produce a set of sketches, with accompanying notes and example locations, to explain the following fault structures: step faults, rift valley, horst, wrench fault.

7 Produce a set of sketches to illustrate the outcrop patterns that might occur at a dip fault and a strike fault.

8 Give a series of annotated diagrams illustrating the principal **concordant** and **discordant** igneous rock formations.

Give a series of diagrams to show how the following features are indicated on a published (Institute of Geological Sciences) maps: dip, overturned strata, horizontal strata, fault, coal seam, borehole, mineral vein, alluvium, landslip, unconfirmed boundaries.

5 Geology of groundwater

5.1 HYDROLOGICAL CYCLE

The outer envelope of the Earth consists of bodies of solid, liquid and gas which are referred to as:

The **Lithosphere** The total body of solid outer crustal rocks.
The **Hydrosphere** The total body of water in oceans, seas, lakes, rivers, etc.
The **Atmosphere** The total body of gas and vapour surrounding the Earth.

The study of the Earth's water, with particular reference to its properties and movement is called **hydrology**, while the study of water in relation to rock masses is called **hydrogeology**. The movement of water is also considered in the allied science of **oceanography** and **meteorology**.

A very large proportion of the Earth's total water occurs in oceans and seas (*Fig 5.1*) a small amount occurs as snow or ice (solid), a smaller amount as land-based water and an even smaller amount in the atmosphere (water vapour). A simple concept of the overall mechanism of water movement is described in the **hydrological cycle** (*Fig 5.2*).

Precipitation is the mechanism whereby water is released from bodies of water vapour (clouds) to fall either, back into oceans and seas, or on to land surfaces as rain

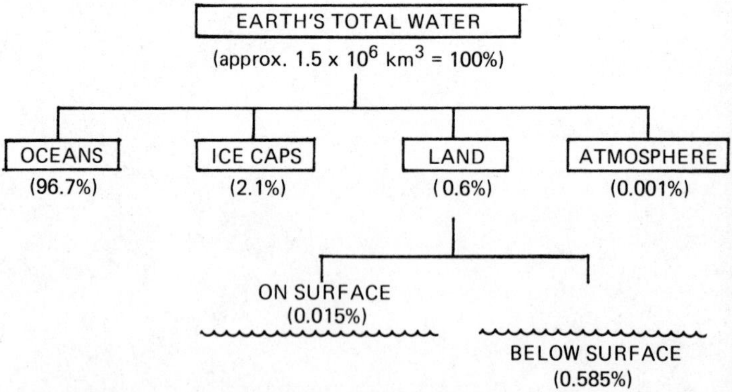

Fig 5.1 Distribution of the Earth's water

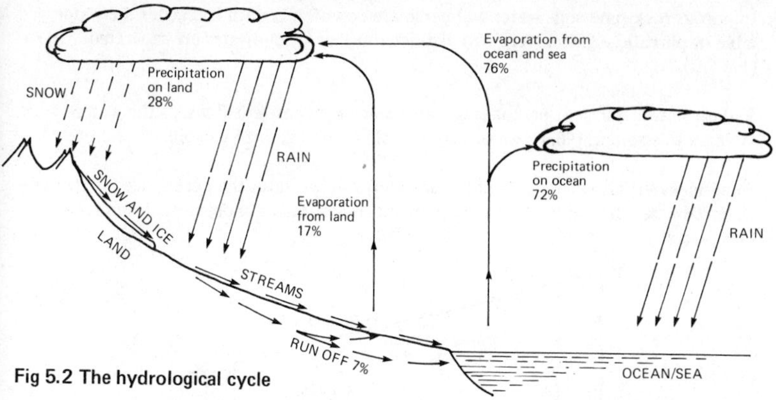

Fig 5.2 The hydrological cycle

snow, hail, sleet, fog, dew, etc. Only about 25% of the total precipitation from the atmosphere falls on to land; some of this is **evaporated** back into the atmosphere, some **runs off** over the surface back to the oceans/seas and some **infiltrates** into the ground.

Evaporation occurs when water vapour is formed, i.e. from:

bodies of water: surfaces of oceans, seas, lakes, reservoirs, rivers, etc.; also from falling rain.
bodies of snow and ice: surfaces of snowfields, glaciers, etc.
vegetation: plants draw water from the ground and evaporate it from their leaves.
land: land surfaces, when warmed, will give off water vapour.
industry/domestic areas: for example, vapour from cooling water, also the burning of hydrocarbon fuels gives off water vapour.

Run-off occurs when the build-up of surface water exceeds the rate at which it can seep into the ground, thus streams and rivers are formed which conduct run-off water back to the oceans/seas.

Infiltration occurs when water seeps into the surface rocks or soils and so becomes groundwater; the process is controlled by
(a) *surface factors*: Slope and nature of ground surface (i.e. porosity, vegetation), and also temperature and rainfall.
(b) *transmission factors*: Porosity and permeability of rocks and/or soils.
(c) *storage factors*: Structural features, e.g. bedding, folds, faults; inflow and outflow streams.

Rainfall is measured in mm of relative accumulated depth. The average rainfall for Great Britain as a whole is about 1 m/year; in England it is about 850 mm/year. There is, however, considerable variation across the country with some places receiving over 3 m/year (e.g. Borrowdale 3.5 m/year) and others receiving less than 200 mm/year. In very dry deserts, the rainfall may be less than 50 mm/year, but in very wet areas, such as the Himalayas, it may be over 10m/year.

5.2 GROUNDWATER ZONES

The term **infiltration** is used to describe the movement of water from the surface into the ground; the subsequent movement of groundwater is described as **percolation**.

In porous rocks and soils water will percolate downwards until it reaches the **water table** or **phreatic surface.** Here two distinct zones of groundwater are separated. (*Fig 5.3*).

Vadose water: transient percolating water moving downwards towards the water table; in this zone **capillary water** may be held around grains by capillary suction.

Phreatic water: also called **gravitational water**; this saturates the pore-space below the water table and has an internal pore-pressure greater than atmospheric; it tends to flow laterally.

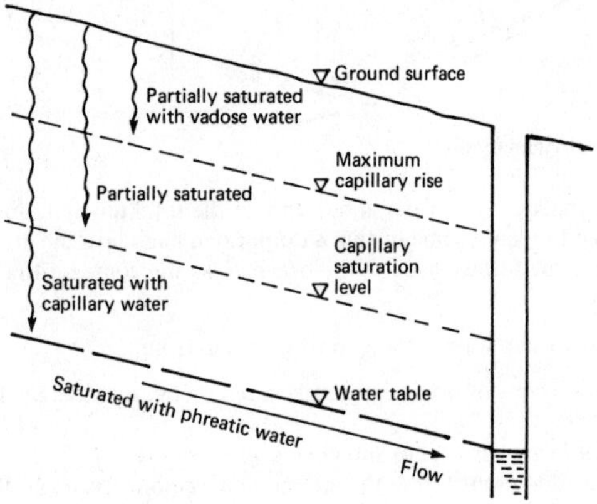

Fig 5.3 Groundwater zones

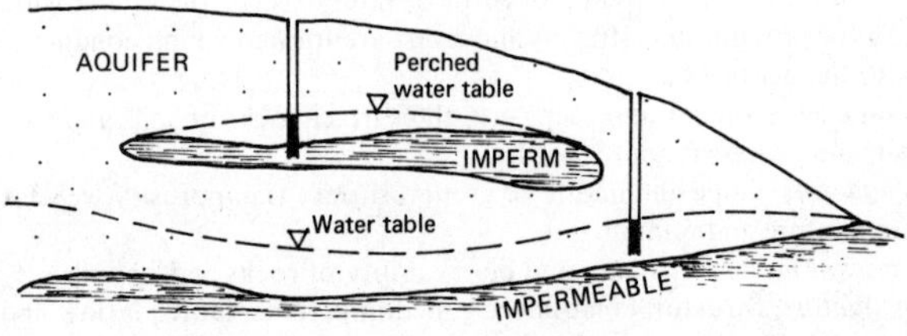

Fig 5.4 Normal and perched water tables

WATER TABLE OR PHREATIC SURFACE

The **water table** or **phreatic surface** is the upper limit of gravitational water and corresponds to the free surface of an open body of water (*Fig 5.3*). In a saturated rock or soil mass, the water table is the level at which the porewater pressure is equal to atmospheric pressure. Water in boreholes and excavations will stand at the water table level. Where groundwater lies above isolated bodies of material of low permeability such as clay, a **perched water table** is said to occur (*Fig 5.4*).

CAPILLARY WATER

In fine-grained soils and particularly in fine flaky soils such as clays, which have very large surface areas, water is held around individual grains by **suction**. This phenomenon, called **capillary suction**, arises because these flat flaky surfaces are negatively-charged and therefore attract the positive ends of bipolar water molecules.

Capillary water may be drawn up above the water table to heights of several metres in clay soils, up to about 1 m in fine sands, but up to only a few millimetres in coarse sands and gravels.

5.3 MOVEMENT OF GROUNDWATER

Water will percolate through rock or soil masses which are either sufficiently permeable or contain discontinuities.

Permeable rocks and soils will allow the passage of water through their unbroken mass; their pores are both large and interconnected.

Impermeable materials will not allow the passage of water; rocks and soil with only micropores or with non-connective pores are virtually impermeable under conditions of constant stress.

Pervious rocks and soils are either permeable (i.e. highly porous) or they contain discontinuities (e.g. bedding planes, joints, fissures, faults, etc.) through which water can flow. For example, sands, gravels, limestones, sandstones and jointed igneous rocks.

Impervious rocks and soils have very low permeability and do not contain discontinuities, thus groundwater will not pass through them. For example, clays, shales, unjointed lavas or intrusive sheets.

A stratum of permeable or pervious rock or soil capable of allowing the passage of water is called an **aquifer**; an **aquiclude** is an impervious stratum.

UNCONFINED AQUIFERS

An aquifer is said to be **unconfined** when the ground surface forms its upper boundary (*Fig 5.5*). The area of land drained by a particular river of stream is termed the **catchment area**; the dividing line between two adjacent catchment areas is called a **watershed**.

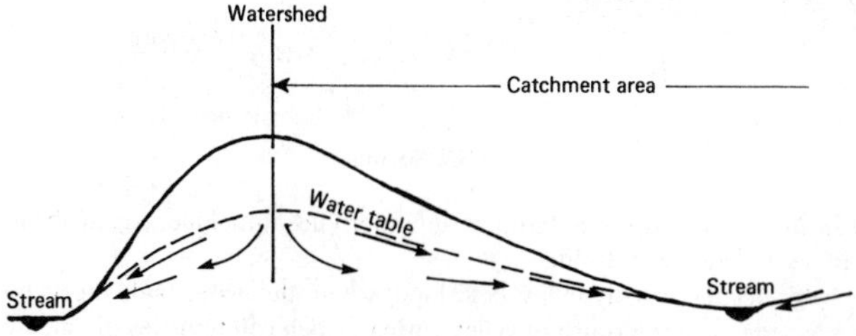

Fig 5.5 Unconfined aquifer

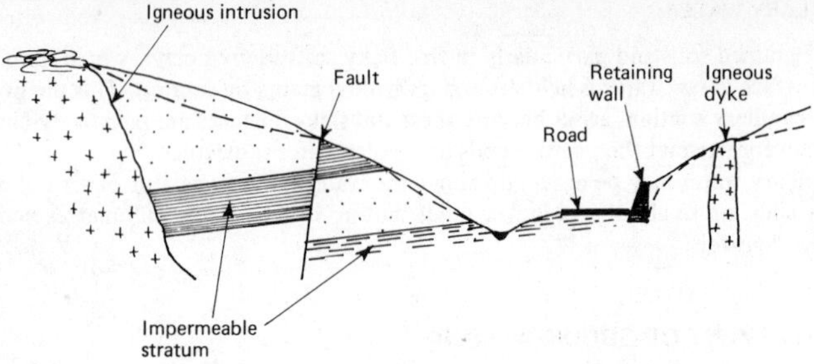

Fig 5.6 Impermeable boundaries affecting the flow in aquifers

When the permeability of a rock or soil mass is the same in all directions it is said to be **isotropic**; this may be the case in non-fissured porous materials, such as sands and gravels. Stratified deposits wherein the layers have different vertical and horizontal permeabilities are termed **anisotropic**; the presence of fissures or other discontinuities will also produce anisotropic conditions. In **heterogeneous** aquifers the permeability varies both vertically and laterally, e.g. alluvial and fluvio-glacial deposits.

The storage capacity and flow characteristics of an aquifer are also affected by its **impermeable boundary** conditions. Impermeable boundaries may be formed by underlying aquicludes, fault junctions with aquicludes, igneous intrusions or man-made structures (*Fig 5.6*).

Groundwater discharges from an aquifer on to the ground surface at a **spring**, the occurrence of which is dependent on the location of impermeable boundaries (*Fig 5.7*).

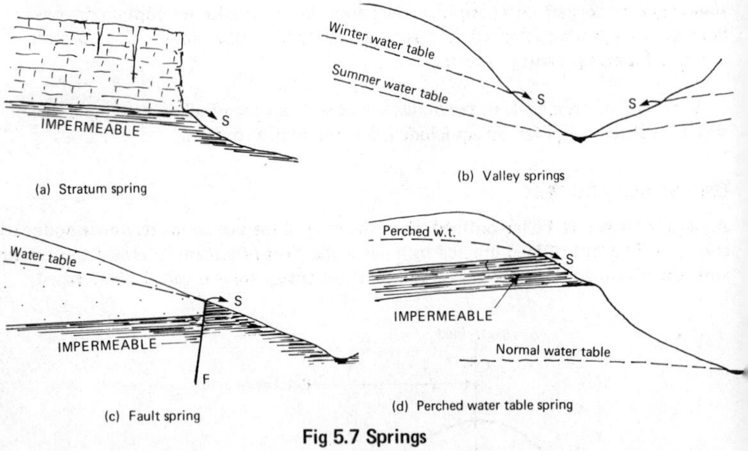

Fig 5.7 Springs

Stratum spring: occurs at the outcrop of an underlying aquiclude, e.g. at the junction of limestones and a clay stratum.

Valley spring: occurs (usually in a valley slope) where the water table meets the ground surface; a *bourne* is an intermittent valley spring, which often moves up and down the valley slope as the water table rises and falls in response to seasonal conditions.

Fault spring: occurs where an aquiclude is faulted to the surface against an aquifer.
Perched water table spring: a variation of a stratum spring often occurring high up the valley slope and above the main water table.

CONFINED AQUIFERS

When an impermeable layer forms its upper boundary, an aquifer is **confined** and so possesses the property of being able to store water under pressure. The groundwater conditions in a confined aquifer are said to be **artesian** and the discharge points called either **artesian springs** or **artesian wells.**

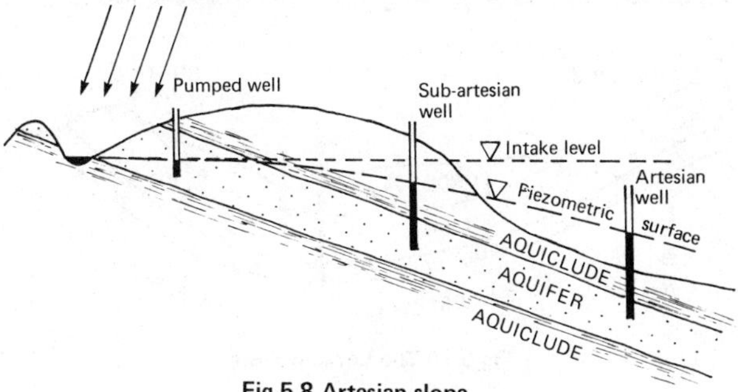

Fig 5.8 Artesian slope

Fig 5.8 shows a cross-section through an **artesian slope**, in which the dipping aquifer is sandwiched between aquicludes. The water-bearing Spilsby Sandstone of Lincolnshire is a typical example of an artesian slope.

If standpipes were to be inserted into the aquifer, the water level within them would rise until it reached the **piezometric surface.** The piezometric head represents the artesian pressure in the aquifer. Three types of well are possible (*Fig 5.8*).

In a:	*the piezometric surface is:*
Normal pumped well,	below upper boundary of aquifer.
Sub-artesian well,	above aquifer, but below ground surface.
Artesian well,	above ground surface.

Artesian springs reach the ground surface under artesian pressure in suitable impermeable boundary conditions, such as along porous faults (*Fig 5.9a*). **Blown-out** or **Blow-out** springs occur where the artesian water pressure is sufficient to punch a hole through the overlying aquiclude (*Fig 5.9b*). This has often occurred during the drilling of boreholes or wells, as for example near Sleaford, Lincolnshire where, following the drilling of a bore-hole in search of coal, water burst through from a limestone aquifer and now flows upward through 30 m of clay to form a small stream.

An **artesian basin** is formed where a synclinally folded aquifer is confined, as for example in the London Basin (*Fig 5.10*), where the Chalk forms an aquifer confined by the London Clay above and by the Gault Clay below. Other examples, are the Paris Basin in France and the Queensland Basin in Australia.

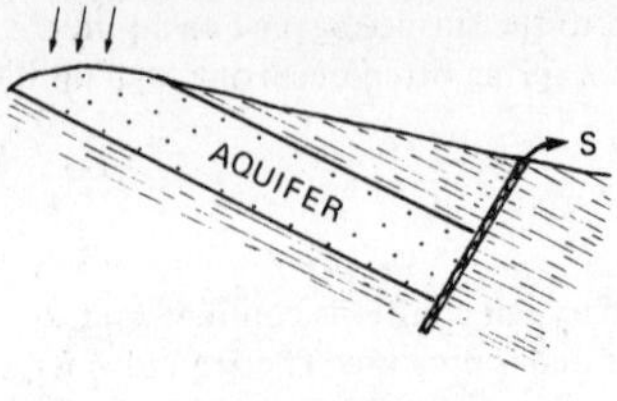

(a) Faulted artesian spring

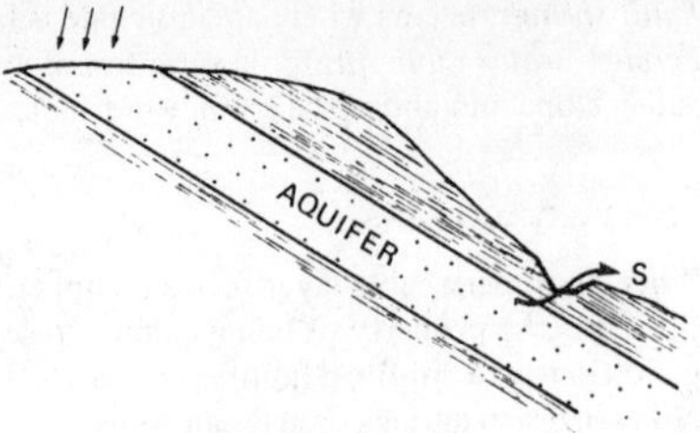

Fig 5.9 Artesian springs

(b) Blown-out artesian spring

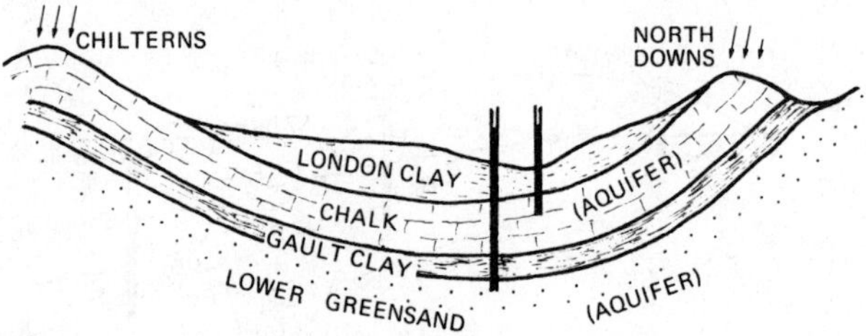

Fig 5.10 The London Basin

5.4 SALINE CONDITIONS

In coastal regions (within about 1 km) the groundwater may be saline due to the presence of sea water. Although fresh water and sea water will mix, they only do so (in a finely porous rock/soil medium) along a relatively narrow interface. The position of the fresh/saline interface may be estimated using the **Ghyben-Hertzberg** balance relationship. This proposes simply that columns of fresh water and sea water will balance each other if their heights are in an inverse ratio of their densities.

The relationship is illustrated in *Fig 5.11*, where the groundwater regime is considered under an (ideal) island, which is surrounded by sea water. If the specific gravity of sea

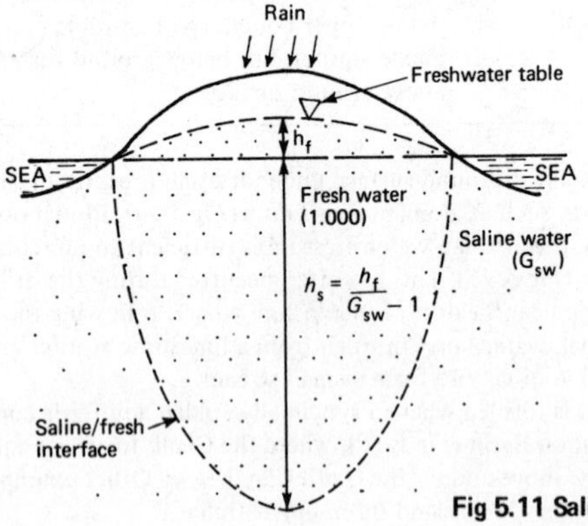

Fig 5.11 Saline/fresh interface

water is G_{SW} and that of fresh water 1.000, then under static conditions, each unit head of sea water will be balanced by G_{SW} units of fresh water. Suppose the height of the fresh water table above sea level is h m, then the depth to the fresh/saline interface at this point will be given by

$$d_{f/s} = \frac{h}{G_{SW} - 1}$$

For example, if the specific gravity of sea water is 1.025 (about average for the Atlantic or North Sea), then the depth to the interface will be forty times the height to the fresh water table.

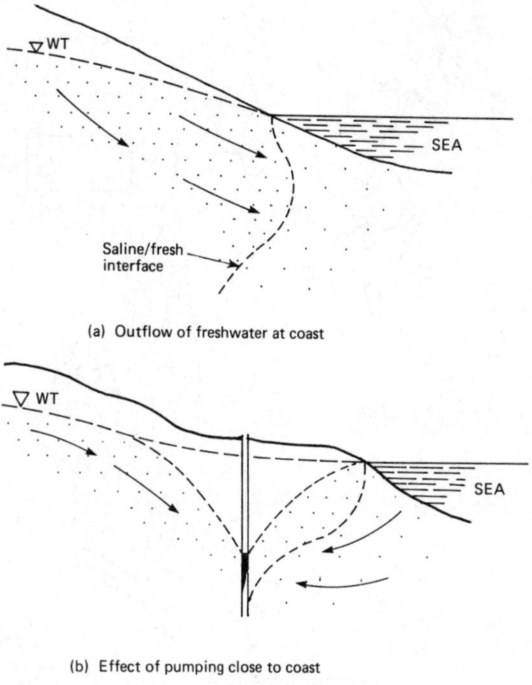

Fig 5.12 Distortions of saline/fresh interface

This relationship provides a simple estimate for static conditions, but becomes less reliable as groundwater flow increases. Where the freshwater flow is directed seawards, the saline/fresh interface will also be pushed seawards and deeper (*Fig 5.12a*). Pumping wells sunk close to the coast will however pull the interface inland and may, as a result, become contaminated with saline water (*Fig 5.12b*). Such conditions apply also to land adjacent to tidal rivers.

5.5 SOURCES OF WATER SUPPLY

Water supply sources are derived from the overall land drainage system by modifying natural storage and flow processes, for example, by constructing dams and pipelines,

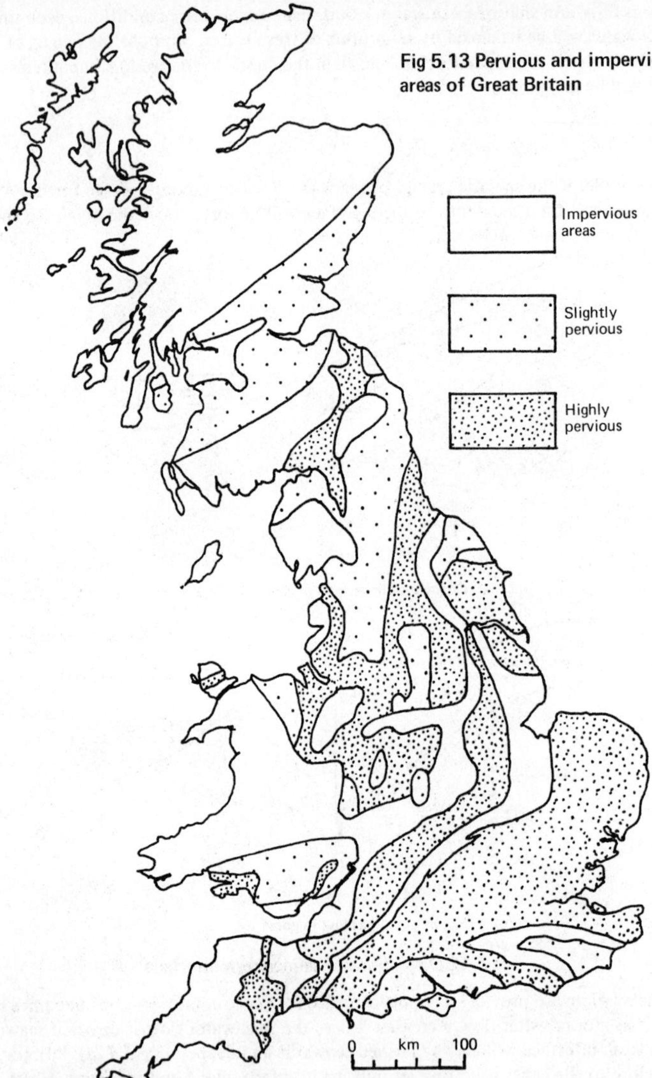

Fig 5.13 Pervious and impervious areas of Great Britain

forming reservoirs, or drilling wells. In areas of pervious rocks the source is underground and in impervious areas it is surface run-off.

It is in some respects fortunate that those areas of the British Isles having the lowest annual rainfall are among the most pervious, while in areas of high rainfall impervious rocks occur. *Fig 5.13* shows a map of England, Wales and Scotland divided very approximately into pervious and impervious areas. The lowest annual rainfalls are recorded in the east and south east and the highest in N.W. Scotland, Wales, the Lake District and S.W. England. Thus the main underground water sources are concentrated in the 'driest' parts of the country.

UNDERGROUND SOURCES

Underground sources contribute significantly to the water supplied to many large towns and cities. The principal formations are described below, but many other small local sources remain in active use.

Carboniferous Limestone: Derbyshire, Yorkshire, North and South Wales and the Mendips (supplying Bristol).
Coal Measure Sandstones: South Wales, Coventry.
Millstone Grit: Eastern Pennines.
Bunter Sandstone: Nottingham, Wolverhampton.
Keuper Sandstone: Warwickshire, Worcestershire, Lancashire.
Magnesium Limestone: Durham and Northumberland.
Jurassic rocks (Great Oolite and Inferior Oolite): from Yorkshire to Dorset, including parts of Lincolnshire, Northants, Bedfordshire and Wiltshire.
Chalk: from Yorkshire to Dorset, including Lincolnshire, the Chilterns, North and South Downs and Salisbury Plain; supplying for example London (15% of it total), S.E. coastal towns from Margate to Dover to Brighton, Portsmouth, Southampton and Bournemouth.
Lower Greensand: Cambridgeshire.
Eocene sands: Cambridgeshire.

SURFACE RUN-OFF SOURCES

The bulk of water supplied to large towns and cities is taken from either rivers, lakes or reservoirs. Natural transportation via streams and rivers is augmented, over considerable distances in some cases, by pipelines and pumping stations. Some principal examples are given below:

River supplies: London, Southampton, Teeside, Tynside.
Lake supplies: Glasgow, Manchester.
Reservoir supplies: Birmingham, Liverpool, Manchester, Sheffield, Plymouth, Torquay.

TEST EXERCISES

COMPLETION QUESTIONS (answers given on page 206)

Complete the following statements by inserting the most appropriate words in the spaces indicated.

1 The study of the movement of the Earth's water is called . . . .

2 The movement and circulation of water between ocean, land and atmosphere is described by the . . . . cycle.

3 The process whereby water is released from clouds is called . . . . ; in returning to a vapour, water is said to . . . .

4 As water seeps into the ground . . . . occurs.

5 Water passing over land surfaces is called . . . .

6 State three factors which control the seepage of water into the ground: . . . ., ..., ...

7 Water percolating downwards is termed . . . . water .

8 The water table represents upper boundary of . . . . water in the ground; another term for the water table is . . . . surface.

9 The pressure in the water at the water table is equal to . . . .

10 When a local body of groundwater lies above the normal water table a . . . . is formed.

11 In fine-grained soils, water is held around the particles by . . . .

12 The ability of soil and rock masses to allow through passage of water is described by the following terms:
(a) will not allow passage of water: . . . .
(b) passage through unbroken mass: . . . .
(c) passage through discontinuities: . . . .

13 An aquifer . . . . allows the passage of water, an aquiclude . . . .

14 A particular river or stream drains a . . . . area; the dividing line between . . . . areas is termed a . . . .

15 When the permeability of a rock or soil mass is the same in all directions it is said to be . . . . ; stratified deposits are usually . . . .

16 At a spring . . . . comes to the surface.

17 An intermittent valley spring is called a . . . .

18 In a confined aquifer the conditions are said to be . . . . when the . . . . surface lies above the upper boundary of the aquifer.

19 Discharge from confined aquifers may often be from . . . . springs or . . . . wells.

20 The London basin is an . . . . basin in which the Chalk is the principal . . . .

21 The Ghyben-Hertzberg relationship is used to estimate the position of the . . . . interface.

22 The density of seawater in the North Sea is approximately . . . .

23 The highest annual rainfall in Great Britain occurs mainly in . . . .

24 Name three rock formations in Great Britain that provide major sources of water supply: . . . . , ... , ...

ESSAY QUESTIONS

1 Draw a diagram to represent the hydrological cycle and explain the additions and losses occurring in the bodies of water on and in the land.

2 Discuss the factors which control the infiltration, percolation and storage of groundwater.

3 Distinguish between vadose and phreatic water, and also between water table and piezometric surface.

4 Explain the phenomenon of capillary water in the ground and discuss the problems that may arise in connection with this from an engineering point of view.

5 With the aid of sketches, describe the geological conditions which control the emergence of groundwater from unconfined aquifers.

6 Explain the occurrence of artesian conditions in certain aquifers, giving examples and sketches to illustrate the points made.

7 Discuss the pollution of groundwater that can occur both naturally and due to the activities of man.

8 Give a brief account of the groundwater problems that may be encountered on a large land-based construction project.

6 History, scope and application of soil mechanics

6.1 A HISTORICAL REVIEW

Ever since man has been engaged in putting up buildings, some interest has been focussed on the unsuitability of the ground. Very often (even today) evidence of unsuitability only appears when a building failure occurs. Historically, the first studies of ground conditions were mainly prompted by failures in important buildings.

The importance of **sound** and **safe** construction, however, was recognised long before any form of analysis or predictive design method was thought of. In fact, the earliest known **Building Regulations** were included in the Code of Hammurabi, a ruler of Babylon in the 17th Century B.C. (*Fig 6.1*). While today's codes of practice and statutory regulations are more detailed, it would seem the penalties for infringement are far less severe.

Up to the 17th century (AD), the only interest in soil or rock conditions was that shown by military engineers in their designs for fortifications. Site investigation, if at all considered, was perfunctory. The first record of a deep exploratory borehole appears with description of the construction of the Stadt House in Amsterdam, where it seems a hole 71 m deep was drilled. In 1666, Sir Christopher Wren had trial pits dug down to 10 m deep prior to the building of St. Paul's Cathedral in London. In spite of this, parts of the foundations were found in 1850 to be unstable. Investigation were ordered in 1743 by Pope Benedict XIV following the occurrence of cracks in St. Peter's Cathedral, Rome. Similar circumstances arose at Strasbourg Cathedral in 1750.

It may have been such incidents that prompted a French physicist, Charles August de Coulomb, to produce the first mathematically-based earth pressure theories. His paper to the Royal Academy in Paris in 1776 heralded the birth of soil mechanics. Coulomb's theories were taken up by several other physicists and mathematicians throughout the 19th century. In 1856, H. Darcy produced experimental results concerning the permeability of sands, and in 1857 W.J.M. Rankine presented a paper to the Royal Society of London entitled *On The Stability of Loose Earth*. In 1862, Rankine published a manual of civil engineering, in which methods were proposed for calculating bearing capacities of footings and lateral earth pressures.

The first scientifically based field investigations were probably those undertaken in 1846 by the Frenchman, Alexandre Collin, in the course of his studies of slip failures in clay cuttings, embankments and earth dams. Another Frenchman, J. Boussinesq, in 1885, made a major contribution to the analysis of stresses in elastic bodies by publishing sets of equations.

If a contractor builds a house for a man, this man shall give the contractor two shekels of silver per ser (unit of weight) as recompense.

If a contractor builds a house for a man and does not build it strong enough, and the house which he builds collapses and causes the death of the house owner, then the contractor shall be put to death.

If it causes the death of the son of the owner, then the son of the contractor shall be put to death.

If it causes the death of a slave of the owner, then he (the contractor) shall give the owner a slave of equal value.

If it destroys property, he (the contractor) shall replace what has been destroyed, and because he did not build the house strong enough and it collapsed, he shall rebuild the house at his own expense.

If a contractor builds a house for a man and does not build it so that it stands ordinary wear and a wall collapses, then he shall reinforce the wall at his own expense.

Fig 6.1 The code of Hammurabi (Engraved on a stone column in the Louvre, Paris)

Just after the turn of the century, the study of soil physics spread to other European countries. Notably, to Sweden, where much work was done on the consistency of soils and the behaviour of slopes by such men as Atterberg, Petterson and Fellenius. In 1922, following a series of failures in railway embankments, the Swedish Geotechnical Commission was formed to study these problems. In Germany, following similar problems encountered during the construction of the Kiel Canal, Professor Krey commenced investigations into the stability of slopes and retaining walls. In Great Britain, A.L. Bell, who had been following the work of Rankine, presented a paper to the Institution of Civil Engineers in which he proposed an analytical method for the calculation of lateral pressure in clays.

The now legendary figure of Karl Terzaghi first made his presence felt in the new science of soil mechanics when he published his theory of consolidation in his native Vienna in 1925. In the same year he emigrated to the USA and joined the Massachusetts Institute of Technology, where he later became Professor of Soil Mechanics. The theory Terzaghi proposed was based on a simple mechanism and, worked through to an elegant mathematical solution, it provided a practical method for estimating the time required for consolidation settlement to take place. Mathematical analysis, conceptual theory and practical measurement (involving both laboratory and field observations) were brought together to provide substance for his ideas.

A new era of understanding in soil mechanics followed this first publication. Other impressive works followed, as for example his paper read to the American

Society of Civil Engineers entitled *The Science of Foundations – It's Present and Future*. His now famous *Theoretical Soil Mechanics,* the first soil mechanics textbook, was published in 1943, and in 1948, with an American engineer Ralph Peck, he was co-author of *Soil Mechanics in Engineering Practice*, another work of undoubted authority and value. So important has been Karl Terzaghi's contribution to ground engineering that he has keen dubbed internationally as the 'father of soil mechanics'.

In 1939, Dr. Terzaghi delivered a lecture to the Institution of Civil Engineers on the subject of soil mechanics, and in 1951 he presented a paper to the Building Research Congress entitled *Soil Studies and the Construction of Foundations*. In this latter paper one of his concluding remarks was:

> "Unfortunately, in practice, clear and simple soil conditions are rather uncommon; and if the soil conditions are complex, the advanced state of soil mechanics is of no avail, unless the engineer in charge of the soil exploration is fully aware of the virtues and deficiencies of the different techniques, and capable of adapting these techniques to the local soil conditions and exigencies of the job."

In other words, the science and art of soil mechanics is as good or as bad as those practising it. A homily that has not changed and one well worth remembering today.

The academic respectability and practical acceptance of soil mechanics was establish in 1932 when the first International Conference was held at Harvard University. Papers were presented by scientists and engineers from many countries. It is interesting to not however, that only one British delegate attended and only two British papers read.

Some of the first soil mechanics research in Great Britain was carried out by Profess C.F. Jenkin at Oxford University. His topic was lateral pressure in granular materials and this he continued after joining the Building Research Station at Garston in 1929. The soil mechanics section of the BRS was established in 1935, following work by Drs. L.F. Cooling and A.W. Skempton on classification tests and correlations for British soils.

The importance of soil mechanics from a practical engineering standpoint was underlined in a report produced by the BRS on the failure in 1937 of part of the Chingford Reservoir. The expertise and authority of the BRS team was emphasised by Karl Terzaghi. He had been brought over from Paris especially and agreed fully with the findings contained in the report. The railway companies at this time recognised the value of soil mechanics in the design of cuttings and embankments, and in 1938 began to contribute financially.

Also in 1938, the science and art of soil mechanics *arrived* in the academic sense, when the first courses of lectures were given at British Universities. By 1940, the teaching of soil mechanics had commenced at University College, London, and at the Universities of Durham, Glasgow and Sheffield. Although the outbreak of war in 1939 curtailed some of the work and the development of teaching, urgent research into practical applications was developing at the BRS and the Road Research Laboratory. As well as work on site investigation and testing, studies were carried out on a wide range of failures in earth dams, retaining walls, dock and harbour walls, factory buildings, cold stores, storage structures and drainage systems.

In the USA also, the war produced an impetus to develop practical methods, particularly in road and airfield design and construction. It was from such work that Albert Casagrande proposed a classification system for soils incorporating a comprehensive arrangement of sub-groups; this still forms the basis of the present British Soil Classification System given in BS 5930: 1981.

A second international conference on soil mechanics was held in Rotterdam in 1948, and similar conferences have since followed at intervals of four or five years. The first British textbook, *The Mechanics of Engineering Soils*, was published in 1949. The authors P.L. Capper and W. Fisher-Cassie were among the few lecturers in the subject at that time, respectively at University College and Newcastle.

In 1943, a committee was convened to prepare codes of practice relating to soil and ground construction Representatives were initially included from the Institution of Civil Engineers and the Ministry of Public Works, but in 1949 a joint committee was formed, which also included members from the Institutions of Structural Engineers, Municipal Engineers and Water Engineers.

The first Code of Practice was published in 1950 on *Site Investigations*, other codes that followed were *Earthworks, Foundations, Earth Retaining Structures* and *Drainage*. Codes of Practice are now issued by the British Standards Institution; for example a new edition of *Site Investigations (BS 5930)* was published in 1981, and revisions of some of the others are expected in 1982–84. Another important British Standard, *Methods of testing soils for civil engineering purposes* (BS 1377) was first published in 1948; the present edition published in 1975 is due for revision in 1982.

In the 1950s and 1960s, the understanding of soil behaviour increased dramatically, particularly with regard to stress-strain relationships and the importance of effective stress analysis. Important contributions were made by A.W. Skempton, A.W. Bishop and others relating to the properties of clays and the stability of clay slopes.

Major developments also took place relating to deep bored piles, large diameter bored piles, diaphragm walls, deep excavations and foundations in poor ground. Instrumentation was improved, with the introduction of electrical strain gauges, settlement gauges, inclinometers and piezometers. Computational methods were revolutionised by the introduction of computers, making possible the use of complex numerical techniques, such as finite element analysis.

In the 1970s, geotechnical engineering, including site investigations, was now seen as a substantial or even separate branch of civil engineering. A large number of companies were able to offer a wide range of design, testing or contracting services. Considerable emphasis was centred on *in-situ* measurement and testing, and improved instrumentation brought about renewed interest in observational techniques, especially in connection with settlement. Theoretical work included the consideration of the long-term residual strength of clays and the notion of critical states. The importance of the structure and fabric of soil in-situ and its influence on measured parameters was a recurring theme.

At the beginning of 1983, much work continues on improving and developing techniques: analytical, empirical and observational. Following the inclusion of probability theory in the design of superstructure (e.g. RC and steel frameworks), some interest is being shown in this in foundation and earthwork design.

In site investigation, the main problems include more reliable means of interpreting in-situ testing and the development of techniques for off-shore work. In laboratory techniques, major advances are expected in the automatic control and interpretation of routine testing, following the introduction of relatively low-cost micro-computing facilities.

.2 SCOPE AND APPLICATION OF SOIL MECHANICS

The soil mechanics engineer or technician needs to be skilled in a number of disciplines and areas of study, for his is not a compact single science, but rather a complex mixture of several. In attempting to define the scope of the various branches and avenues embraced by the generic term **geotechnics**, of which soil mechanics itself is a substantial part, it is necessary first to establish a viewpoint. From an academic point of view, the main disciplines and associated subject areas are those listed below (in no particular order of importance).

geology	*physics*	*hydraulics*	*mathematics*
geomorphology	mechanics	hydrostatics	numerical methods
geophysics	statics	hydrology	computing
geochemistry			probability theory

In addition, certain human intelligence attributes need to be acquired, such as good observational ability, coherent thinking and so-called 'common sense'. From an educational point of view, the division of knowledge into content areas may serve to indicate how teaching/learning processes may be devised:

Knowledge of materials – origin, occurrence, formation, properties, behaviour, endura
Knowledge of principles – mathematical and mechanical concepts.
– theoretical behaviour systems.
Knowledge of criteria – measures or parameters of behaviour and properties.
– laboratory and field measurement.
– limits, safety factors.
Knowedge of problems – occurrence of problems: loading, deformation.
– effects of time; change, stability.
– practical controls: safety, economic, operational.
Knowledge of methods – methods and systems in design and construction.
– problem-solving strategies.
Knowledge of consequences – cause and effect; constancy and changeabiiity; prediction.

It is easy to divide the study of soils in terms of their engineering properties. Likewise, engineering problems may be grouped according to the basic mechanisms involved. Some connections then appear which link certain properties with certain problems. A very simple set of connections is shown in *Fig 6.2*. Over-simplified and over-generalised relationships such as this are not specifically productive, except to indicate the types of relationships that exist. Problem-solving strategies are more valuable when they are related to limited problem areas.

The following examples might be used in connection with a foundation design problem. **Prompt-questions** are used here to define the design process, although other tactics might be equally effective.

Prompt question	*Related design components*
1 What loading is imposed?	Intensity or distribution of load.
2 How is this transmitted through the soil?	Stress distribution, boundary conditions; elastic or plastic.
3 What effects will this have on the soil?	Development of shear stress, compression and volum change.
4 What are the limiting conditions?	Elastic or plastic equilibrium; design criteria.

5	What effects will be experienced by the structures?	Settlement, distortion, secondary stresses.
6	Can all the criteria be met?	Economic, operational, safety.
7	What will happen in the future?	Prediction of changes, long-term strength and deformatio

It should be clearly understood that every problem is unique; many may be similar, but none identical. A good grasp of principles and a sound knowledge of properties and methods are basic requirements in the education and training of soil engineers or

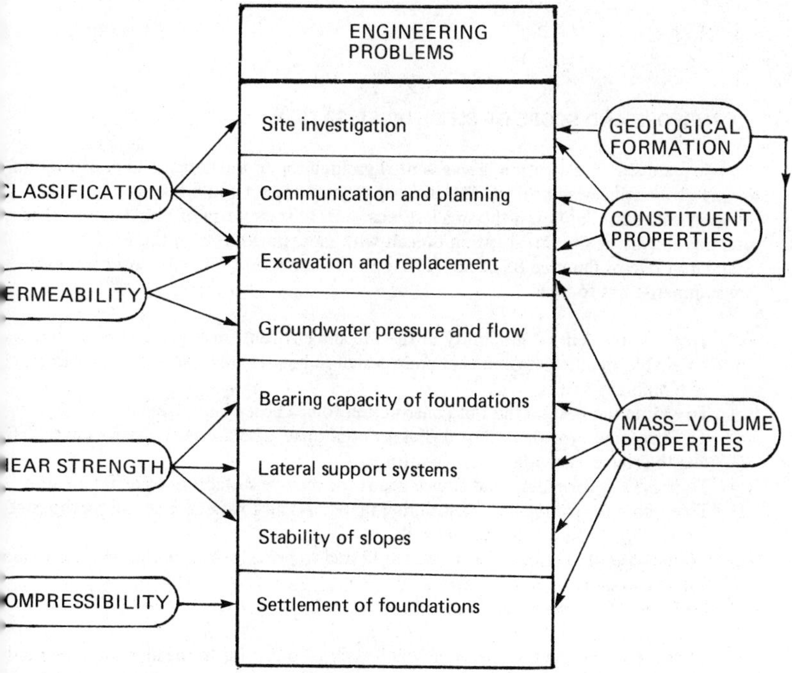

Fig 6.2 Principal soil properties featuring in ground engineering problems

technicians. However, it is the undefinable attributes that separate the good from the indifferent, the capable from the muddled.

The possession of mere mathematical or mechanical knowledge will not in itself confer ability in solving problems as complex as those associated with ground engineering. Knowledge of the materials themselves, of the rocks and soils, in intimate detail and with a clear insight into their behavioural properties is the real key to successful and able study in soil mechanics.

It is not pressing the point too severely to remind the reader once again of Dr. Terzaghi's remarks to the Building Research Congress (see page 82). Whatever has happened before, and however elegant the analysis, it is **this** soil on **this** site behaving in **this** way that is going to affect **this** building.

7 Site investigation and exploration

7.1 PURPOSE AND SCOPE OF SITE INVESTIGATION

An adequate site investigation is an essential preliminary to the design and construction of any civil engineering project. The size and type of project will influence the scope of an investigation, but even the smallest jobs warrant some form of investigation. The practice and art of site investigation is dealt with most thoroughly in the British Standard Code of Practice BS 5930 : 1981, *Site Investigation*. The primary objectives are summarised as follows:

(a) To assess the general suitability of the site and environs for the proposed works.
(b) To enable an adequate and economic design to be prepared, including the design of temporary works.
(c) To enable the best method of construction to be chosen and planned.
(d) To forsee and provide against difficulties that may arise during and after constructi due to ground and other local conditions.
(e) To provide information and advice about the relative suitability of different sites.
(f) To explore for sources of construction materials and to select sites and methods of disposal of waste or surplus materials.
(g) To report on the safety of existing works and to provide information in connection with extensions to existing works.
(h) To investigate cases where failure has occurred.

Most of the work of site investigation will be carried out prior to the start of constructi but in large projects some on-going observations and/or in-situ testing may be necessary. For a moderate to large project a site investigation will proceed in six stages.

1 *Preliminary investigation*: a desk study involving the collection of information and documents: maps, drawings, details of existing or historic development, planning details, location of site and access, services.

2 *General site survey and inspection*: inspection of the site by specialists, such as geologist, land surveyor, soils engineer; also preparation of an overall site plan, giving broad details of topography, geology, access, existing features and works, etc.

3 *Detailed site exploration and sampling*: investigation of detailed geology and soil conditions using surface surveys, trial pits, boreholes, soundings and other methods

as appropriate; survey of drainage and groundwater conditions; survey of existing and adjacent structures; location of buried services and other existing works; provision of samples for further examination and laboratory testing.

4 *Laboratory testing*: tests on disturbed and undisturbed soil samples obtained from the site for classification, quality, permeability, shear strength, compressibility, etc.; tests on rock cores and samples for strength and durability; petrographic examination and chemical analysis.

5 *In-situ testing*: tests on site, either before or during construction, e.g. shear vane, standard penetration, cone penetration, plate bearing, pressuremeter tests; also structure loading tests, such as tests on piles; and post construction monitoring in certain cases, e.g. settlement gauging, slope stability observations.

6 *Reporting*: written report of all procedures and results, which is addressed to client; including site maps and drawings, details of location and geology, borehole logs, tables of results, interpretation of test and survey data, recommendations and advice relating to design and construction.

7.2 METHODS OF GROUND EXPLORATION

The choice of ground exploration method depends on the combined consideration of **four** sets of factors:

1 *Geological factors*: trial pits are only practicable in firm ground; unlined holes may be sunk in clay, but linings will be needed in sands; in hard rocks coring drills will be required; position of the water table may be important.

2 *Topographical factors*: special equipment and methods are required on steeply inclined sites; waterlogged or swampy sites pose special problems.

3 *Information factors*: both designer and contractor require information of particular kinds; number and depth of trial holes must be considered in relation to the layout and size of the proposed structure; specialist information (e.g. seismic data, water flows, joint patterns in rocks) will require special methods.

4 *Economic factors*: both cost and time are important and increase as the extent of exploration increases; careful planning and use of standard methods will reduce cost and time.

TRIAL PITS

Trial pits are preferable to borings in cohesive soils and soft rocks, provided they are not more than 2–3 m deep. A mechanical excavator or even hand-digging will soon expose a succession of strata to easy visual examination. The sides of trial pits require support in most cases and they should be refilled as quickly as possible. Samples can be dug or cut from the bottom and sides. Trial pits are useful for locating buried pipes and services and for groundwater observations.

HEADINGS

Headings (also called **adits**) are excavated almost horizontally, either from steep slopes or cuttings, or from the bottom of shafts. The cost of sinking shafts and driving headings is very high and problems of drainage often arise, so that headings are only used in special situations, such as for mineral surveys or tunneling work.

HAND AUGER

The hand auger (also called a **post-hole** or **Iwan auger**) is a very simple tool used for drilling down to maximum of 5–6 m in soft soils. It usually consists of a 100 mm clay auger, attached by a series of 1 m long extension rods to a crosspiece that may be turned by hand at the surface (*Fig 7.1*).

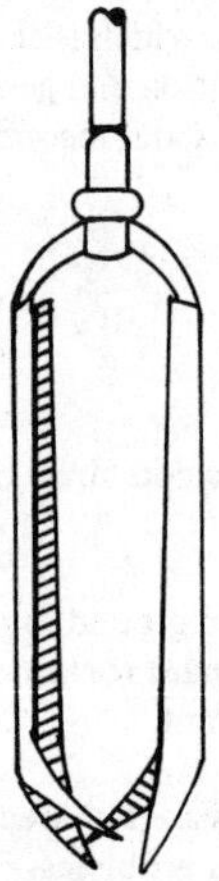

Fig 7.1 Hand auger

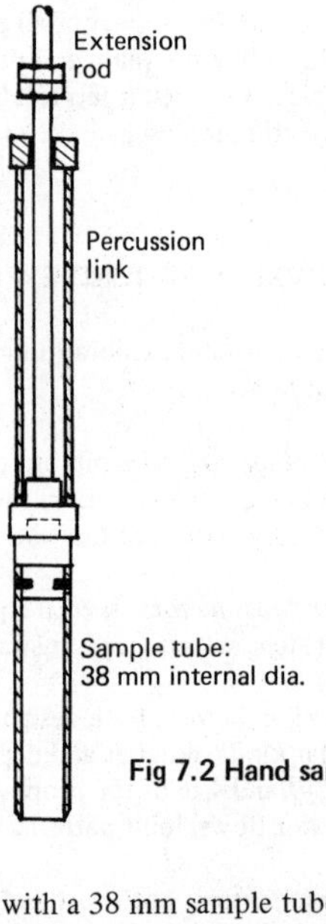

Fig 7.2 Hand sampler

Samples are obtained by replacing the clay auger with a 38 mm sample tube attached to a percussion link (*Fig 7.2*). The sample tube is forced into the soil at the bottom of the hole by raising and dropping the extension-rod/percussion-link assembly. When the tube is full, the cross head is rotated to shear off the soil, and the assembly driven upwards using the percussion link.

SHELL AND BAILER BORING

The shell and bailer method is extensively used, since it is simple and economic in soft to firm soils free of cobbles and stones. It may be used down to depths of 50–60 m.

A derrick, incorporating a power winch is used with a set of drillings tools. The cutting tool, attached to boring rods, is driven by percussion into the soil. When full of soil, the tool is brought to the surface and emptied. Disturbed samples are usually taken at suitable depths.

In clay soils, a **clay cutter** (*Fig 7.3a*) is used, while in sands and gravels, a **sand shell** or **bailer** (*Fig 7.3b*) is used. In the latter, a flap valve ensures that loose or slurried material is retained in the shell as it is raised to the surface. In wet, very soft or loose soils, a length of **casing** must be installed at the surface; steel tubes are knocked or jacked in as drilling proceeds, and may either be drawn out after the hole has been drilled, or left in place to facilitate further observations.

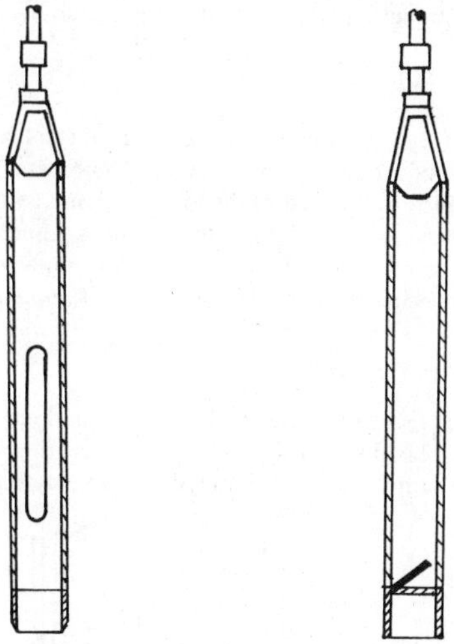

(a) Clay cutter (b) Bailer

Fig 7.3 Percussion drilling tools

ROTARY AUGERS

Power-operated augers, mounted on vehicles, are used in a variety of operations, including site investigations and pile boring. They consist of helical screw augers ranging in diameter from 75 mm to 2 m and capable of sinking holes up to 60 m deep (depending on diameter).

CORE DRILLING

Core drilling is used in hard soils and in rocks. A hollow **coring bit** is attached at the lower end of a **core-barrel** and lowered to the bottom of the hole on boring rods. The assembly is rotated at speeds ranging between 600 and 1200 rev/min, while water is circulated through the coring bit. As an annular cut is made, fragments and slurry are brought to the surface by the circulating water and the core enters the barrel.

After drilling a suitable length (1–3 m), the core-barrel is brought to the surface and the core ejected. The cores are usually stored in slotted boxes and sent to the laboratory for examination and testing.

7.3 SAMPLES AND SAMPLING

Disturbed samples: are collected as drilling or digging proceeds and should be placed in moisture-tight jars or tins, or sealed in plastic bags, together with details of location, depth, etc.; these are mainly required for classification and quality tests.

Undisturbed samples: the **structure**, as well as moisture content, must be preserved as far as is possible; undisturbed samples, sealed in the sample tubes used in boring, together with details of location, depth, etc., are sent direct to the laboratory to be used in tests for shear strength, compressibility or permeability.

OPEN-TUBE SAMPLER

An **open-tube** sampler (*Fig 7.4*) is simply an open-ended steel tube, often threaded at each end and usually of internal diameter 100–105 mm and length 380–450 mm. For drilling, a cutting shoe is attached at the lower end and a sampler head, incorporating a non-return valve, attached at the top. As the sampler is driven into the ground the valve allows air (and water) to escape, but it remains closed as the assembly is raised to the surface, thus retaining the soil sample in the tube.

THIN-WALLED SAMPLER

A **thin-walled** sampler (*Fig 7.5*) is preferred in soft silts and clays which are sensitive to sample disturbance. The lower end is machined to a cutting edge, with an inward reduction in diameter. Samples up to 100 mm diameter may be obtained.

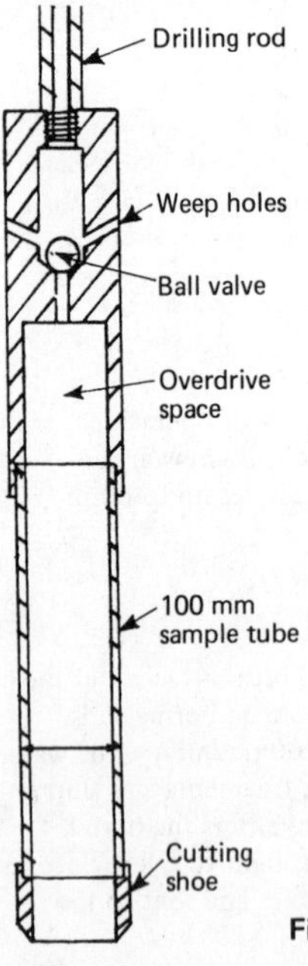

Fig 7.4 Open-tube sampler

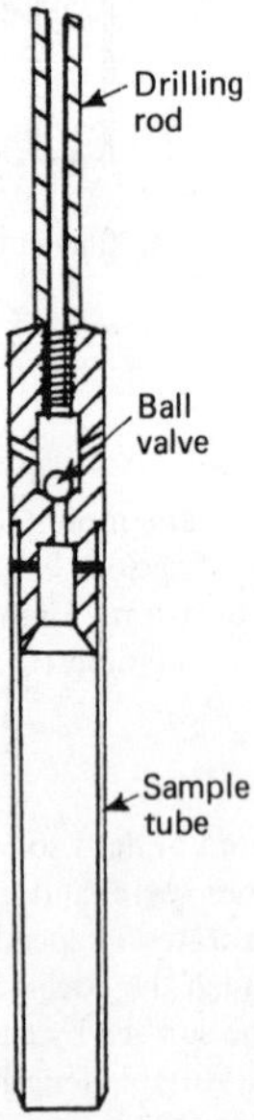

Fig 7.5 Thin-walled sampler

PISTON SAMPLER

For soils of even greater sensitivity, such as very soft alluvial silts and clays, a **piston sampler** (*Fig 7.6*) is required. This consists of a thin-walled tube, fitted with a close-fitting sliding piston that is attached to a shaft passing through hollow boring rods.

With the piston locked at the lower end, the sample tube is held against the soil at the bottom of the hole. The piston is then unlocked and the tube driven into the soil. When the tube is full, the piston is locked again and the sampler brought to the surface. Samples up to 100 mm diameter can be obtained.

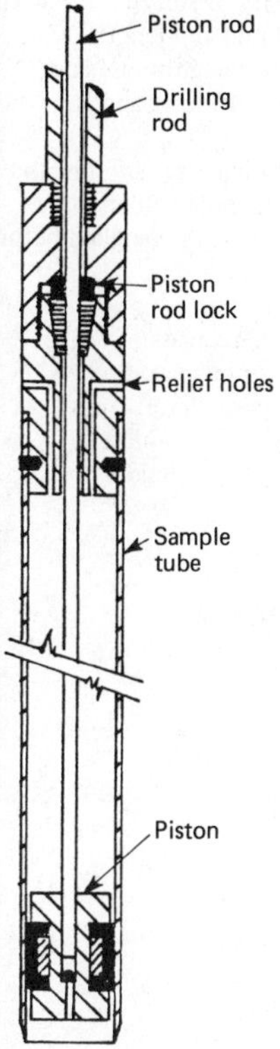

Fig 7.6 Pistol sampler

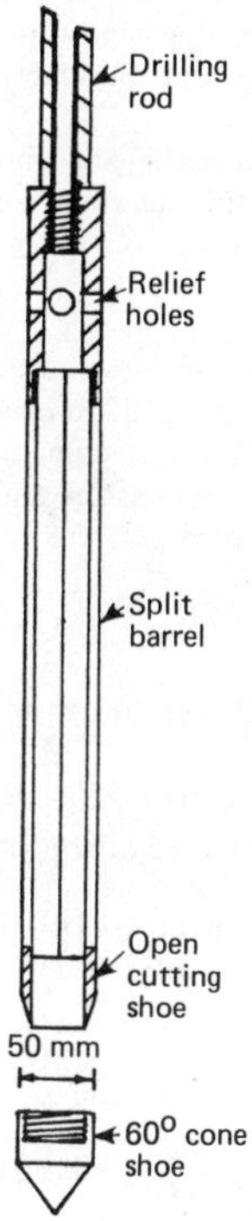

Fig 7.7 Split-barrel sampler

SPLIT-BARREL SAMPLER

The **split-barrel sampler** (*Fig 7.7*) is a dual purpose tool, which may be fitted with a cylindrical cutting shoe for taking samples, or with a solid conical shoe for performing the standard penetration test. The barrel splits into two hemi-cylindrical halves, with the sampler head and shoes attached by screw threads, thus enabling easy removal of the sample.

Since its wall thickness is greater than that of the other tube samplers, its use is restricted to obtaining samples in which disturbance is not so important.

7.4 IN-SITU TESTS

It is often difficult, and sometimes impossible, to obtain good undisturbed samples in sensitive silts and clays. It is also difficult to represent the exact conditions of structure, pore pressure, etc. in the laboratory. *In-situ* tests provide a means of measuring critical properties under actual site conditions.

DENSITY TESTS

The **core-cutter** test is used primarily to determine the bulk density of soil that has been placed by compaction. A steel cylinder of internal diameter 100 mm and length 130 mm and machined to a cutting edge at one end, is driven into the ground using a special rammer. After digging out, the ends of the soil are struck off level and the whole thing weighed.

An alternative method involves digging a hole of approximate diameter 100 mm and depth 150 mm and weighing the excavated soil. The volume of the hole is then determined by pouring into it dry sand from a **sand-pouring cylinder**, which is weighed before and after.

Worked example 7.1 In a core-cutter test a steel cylinder having a mass of 1472g, an internal diameter of 102 mm and a length of 125 mm was rammed into an in-situ soil mass. After removing it and trimming the ends flat, its mass was found to be 3482 g. The moisture content of the soil was later found to be 16.4%. If the specific gravity of the soil is 2.70, determine the bulk and dry densities.

Volume of core cutter $= \frac{\pi}{4} \times 102^2 \times 125 \times 10^{-9} = 1.02 \times 10^{-3}\ \text{m}^3$

Bulk density, $\rho = \dfrac{\text{mass of soil}}{\text{volume of core cutter}}$

$= \dfrac{(3482 - 1472)10^{-6}}{1.02 \times 10^{-3}} = \mathbf{1.97\ Mg/m^3}$

Dry density, $\rho_d = \dfrac{\rho}{1 + m}$ (See section 9.2)

Moisture content, $m = \dfrac{\%\ m/c}{100} = 0.164$

Therefore $\rho_d = \dfrac{1.97}{1.164} = \mathbf{1.69\ Mg/m^3}$

Worked example 7.2 When a sand-pouring cylinder was used in a field density test the mass of sand run into the hole was found to be 1568 g. The mass of soil initially removed from the hole was 1924 g and its moisture content found to be 15.7%. If the density of the pouring sand was 1.65 Mg/m³, calculate the bulk and dry densities of the soil.

Volume of hole = volume of sand poured into it.

$$= \frac{\text{mass of sand}}{\text{density of sand}}$$

$$= \frac{1568 \times 10^{-6}}{1.65} = 0.950 \times 10^{-3}\ \text{m}^3$$

$$\text{Bulk density, } \rho = \frac{\text{Mass of soil removed from hole}}{\text{Volume of hole}}$$

$$= \frac{1924 \times 10^{-6}}{0.950 \times 10^{-3}} = \mathbf{2.025\ Mg/m^3}$$

$$\text{Dry density, } \rho_d = \frac{\rho}{1 + m} = \frac{2.025}{1.157} = \mathbf{1.75\ Mg/m^3}$$

PENETRATION TESTS

The **standard penetration test (SPT)** is a **dynamic** test carried out during the course of drilling a borehole. A standard 50 mm diameter split-barrel sampler is driven into the soil at the bottom of the hole by repeated blows from a drop-hammer of mass 65 kg, falling a distance of 0.76 mm. The sampler is driven into the soil for a total penetration of 450 mm and the number of blows recorded for the last 300 mm. For sands and cohesive soils the cylindrical cutting shoe is fitted and the disturbed samples obtained used for identification. For gravels and soft rocks, the solid conical driving head is fitted.

The **Dutch cone test** is a **static** test, also carried out in a borehole. The cone has an apex angle of 60° and an end area of 1000 mm², and is attached to a rod which is itself protected by an outer sleeve. A measured force is applied to the rod to give a uniform rate of penetration of 15–20 mm/s for about 80 mm. The **cone penetration resistance** is then stated as the force divided by the end area.

VANE TEST

The vane test is designed to measure the *in-situ* undrained shear strength of sensitive cohesive soils, such as soft alluvial clays and silts. A four-bladed cruciform vane is driven into the soil and then rotated at a constant rate until the soil shears. The torque required to shear the soil can be related to its shear strength.

For weak soils ($c_u < 50\ \text{kN/m}^2$) the blade size should be 75 mm wide × 150 mm long, and for slightly stronger soils ($50 < c_u < 100\ \text{kN/m}^2$) the blade size should be 50 mm × 100 mm. The use of the vane test is not recommended for stronger ($c_u \times 100\ \text{kN/m}^2$) or fissured soils.

PERMEABILITY TESTS

The permeability of the soil around a borehole may be estimated by either pumping water out of the hole and observing the rate at which it fills (**rising-head test**), or by pumping water in and observing the rate of fall (**falling head or inflow test**). Another form of inflow test is termed a **constant head test** in which a measured inflow is adjusted to maintain a constant head.

A steady-state flow value is obtained by plotting a flow/time graph and the coefficient of permeability obtained using an empirical equation (several of these exist). Constant head tests are likely to give the most reliable results.

PRESSUREMETER TEST

A device known as a **pressuremeter** is inserted into a suitably sized hole and expanded laterally using compressed gas. Changes in pressure and volume are recorded, from which estimates may be made of the strength and compressibility of the adjacent soil.

7.5 REPORTS

The site investigation report will be addressed to the client who commissioned the work; may be simply a factual account of the work done, with appropriate interpretation, or it may also include comments and recommendations. The whole of the investigation, exploration, testing and interpretation work must be described. Typically the following sections would be included.

1 **Introduction**: a brief summary of the proposed works and the location of the site; the nature of the investigation work, with names and dates.

2 **Description of the site**: general description, including main surface features and access; previous development, including details of existing buildings, services, etc; map of site showing borehole and/or trial pit locations and relevant ordnance datum levels.

3 **Geology of the site**: description of overall geology, related to regional geology of the area; main rock and soil formations; statement of sources and references for further details.

4 **Soil conditions**: detailed account of soil conditions, including types and formations; relevance to proposed works; results of in-situ and laboratory tests; groundwater and drainage observations.

5 **Construction materials**: (where appropriate) details of sources, nature, quantity and significant properties of materials for construction; disposal of waste material and temporary storage of rock or soil.

6 **Engineering interpretation**: worked up results, tables and graphs relating to design, including comments on validity and reliability; where appropriate, recommendations for design and construction; including suggestions for alternative methods; suggestions for further investigation, on-going testing, etc.

7 **Appendices**: collected tables, borehole logs, laboratory test sheets, lists of results, survey records and references.

TEST EXERCISES

COMPLETION QUESTIONS (answers given on page 206)

Complete the following statements by inserting the most appropriate word or words int the spaces indicated.

1 The Code of Practice for site investigation is BS . . . .

2 List the **six** stages of a large site investigation:

. . . .

. . . .

. . . .

3 The choice of a ground exploration method depends on **four** sets of factors, which are: . . . ., ..., ..., ...

4 List **four** of the common methods used in sub-surface exploration:
. . . ., ..., ..., ...

5 If samples are required in which the structure and moisture content are preserved they have to be . . . . samples.

6 In order to obtain good samples in highly sensitive alluvial silts and clays a . . . . sampler should be used.

7 The core-cutter test is used to determine the . . . . of a soil.

8 In the . . . . test a split-barrel sampler is driven into the ground and the number of blows counted.

9 The cone penetration resistance is measured in the . . . . test.

10 The vane test is used to measure in-situ . . . .

11 The pressuremeter test is used to measure . . . . and . . . .

ESSAY QUESTIONS

1 What are the main objectives of site investigations?

2 Discuss the main objectives that would apply to a site investigation in connection with (a) a length of motorway; (b) a compact city site for an office reconstruction.

3 List and explain the stages required in a large site investigation.

4 Suggest the most appropriate site exploration methods in the following cases:
(a) A large grid survey for a light factory development on alluvial silts and sands.
(b) A three-storey school building on deep clays and sandy clays.
(c) Domestic housing on a sloping site on glacial soils.
(d) A multi-storey hotel on bedded limestones and shales.

5 Discuss the need for obtaining undisturbed samples, and compare the relative merits of different sampling methods and devices for this purpose.

6 When are *in-situ* methods likely to be preferred to laboratory tests? Comment on the problems and required precautions associated with *in-situ* testing.

7 Draw up a site investigation report for a fictional project of your own choosing. Under each main heading summarise the (invented) data or information and conclude with typical recommendations or suggestions. (*Note*: Choose a relatively small project on a not-too-difficult site for this exercise).

8 Classification of soils

8.1 PURPOSE AND SCOPE OF SOIL CLASSIFICATION

The principal aim in soil classification is to provide a concise, but meaningful (in an engineering sense), description of the material. The main characteristics that assist classification, and which therefore should be included in a description, fall in two broad categories:

1 **Material Characteristics** These will be determined by physical examination of disturbed samples on site and by laboratory testing: **particle shape and size, grading** (with group symbol where possible), **plasticity indices, mineral composition** (where appropriate) and **colour.**

2 **Structural Characteristics** These will be evident mainly on site, in excavations, trial pits, natural exposures, etc. and sometimes in undisturbed samples: **bedding, discontinuities, moisture content, compactness, weathered condition**; plus the **geological name and age** (where appropriate).

Examples: Firm mottled yellow/brown unfissured CLAY of medium to high plasticity. (Lower Lias Clay)
Medium dense reddish brown silty SAND with sand/gravel lenses. (Flood Plain Alluvium)

A recommended system of field identification and classification is given in *BS 5930 'Site Investigations'* and reproduced here in part as *Table 8.1.*

Where soils are used for constructional purposes, a more quantitative system of description is required. For this purpose, and when more detailed size and grading information is to be conveyed, the use of the *British Soil Classification System for Engineering Purposes (BSCS)* is recommended. The BSCS is set out and explained in BS5930 and summarised here as *Table 8.4.*

Two main groups of inorganic soils are identified:

1 **Coarse-grained** These are retained on a 63 μm sieve and are sub-divided according to particle size.

2 **Fine-grained** These pass through a 63 μm sieve and are sub-divided according to their plasticity.

TABLE 8.1 Field identification and description of soils (Part of Table 6: BS 5930 : 1981)

	Basic soil type	*Particle size, mm*	*Visual identification*	*Particle nature and plasticity*
Very coarse soils	BOULDERS	200	Only seen complete in pits or exposures	Particle shape: Angular Subangular Subrounded Rounded Flat Elongate
	COBBLES	60	Often difficult to recover from bore-holes.	
Coarse soils (over 65% sand and gravel sizes)	GRAVELS	coarse 20 medium 6 fine 2	Easily visible to naked eye; particle shape can be described; grading can be described Well graded: wide range of grain sizes, well distributed. Poorly graded: not well graded. (May be uniform: size of most particles lies between narrow limits; or gap graded: an intermediate size of particle is markedly under-represented.)	
	SANDS	coarse 0.6 medium 0.2 fine 0.06	Visible to naked eye: very little or no cohesion when dry; grading can be described. Well graded: wide range of grain sizes, well distributed. Poorly graded: not well graded. (May be uniform: size of most particles lies between narrow limits; or gap graded: an intermediate size of particle is markedly under-represented.)	Texture: Rough Smooth Polished
Fine soils (over 35% silt and clay sizes)	SILTS	coarse 0.02 medium 0.006 fine 0.002	Only coarse silt barely visible to naked eye; exhibits little plasticity and marked dilatancy; slightly granular or silky to the touch. Disintegrates in water; lumps dry quickly; possess cohesion but can be powdered easily between fingers.	Non-plastic or low plasticity
	CLAYS		Dry lumps can be broken but not powdered between the fingers; they also disintegrate under water but more slowly than silt; smooth to the touch; exhibits plasticity but no dilatancy; sticks to the fingers and dries slowly; shrinks appreciable on drying usually showing cracks. Intermedate and high plasticity clays show these properties to a moderate and high degree, respectively.	Intermediate plasticity (Lean clay) High plasticity (Fat clay)
Organic soils	ORGANIC CLAY, SILT or SAND	Varies	Contains substantial amounts of organic vegetable matter.	
	PEATS	Varies	Predominantly plant remains usually dark brown or black in colour, often with distinctive smell, low bulk density.	

(Reproduced by permission of the British Standards Institution)

TABLE 8.2 Description of composite soil types.

Predominantly coarse-grained	e.g. SAND or GRAVEL
Descriptive term	*% clay or silt*
Slightly clayey or slightly silty	0–5
Clayey or silty	5–10
Very clayey or very silty	10–35
Predominantly fine-grained	e.g. CLAY or SILT
Descriptive term	*% sand or gravel*
Sandy or gravelly	35–65
(non used)	0–35

TABLE 8.3 Description of structural or mass characteristics

Descriptive term	*Field identification*
COARSE-GRAINED SOILS	
Loose	Easily excavated; 50 mm wooden peg easily driven in.
Dense	Pick required for excavation; 50 mm wooden peg hard to drive.
Slightly cemented	Excavated lumps hold together when abraded
Homogeneous	Essentially one type
Heterogeneous	Mixture of types
Stratified	Alternating layers
Weathered	Signs of weakening; concentric layers
FINE-GRAINED SOILS	
Very soft	Exudes between fingers when squeezed.
Soft	Moulded with light finger pressure.
Firm	Moulded with strong finger pressure.
Stiff	Cannot be moulded with fingers; can be indented by thumb.
Very stiff	Can be indented by thumb nail.
Fissured	Breaks into polyhedral fragments.
Intact	Not fissured.
Stratified	Alternating layers.
Homogeneous	Essentially one type.
Weathered	Crumbly or columnar structure.

Each sub-group is identified by a descriptive double-letter symbol: the first letter indicates a soil-type name, while the second and subsequent letters indicate a qualifying characteristic. (See *Table 8.5*)

Examples SW = well-graded SAND
SCP = poorly-graded clayey SAND
CHO = organic CLAY of high plasticity

8.2 RAPID METHODS OF CLASSIFICATION

Rapid methods of classification are appropriate for identification in the field, or where laboratory facilities are not available. A combination of sound judgement (based on experience) and simple tests is used in conjunction with *Tables 8.1–8.3*. If soil sub-group symbols are quoted they should be written enclosed in brackets to indicate that they are based on a rapid (as opposed to laboratory) method.

Particle size Gravel sizes (> 2 mm) are apparent visually; sands (< 2 mm) tend to cling together when damp and they feel gritty between the fingers; silts (< 0.06 mm) feel abrasive, but not gritty; clays (< 2 μm) feel greasy.

Grading The *grading* of a soil refers to the distribution of particle sizes: a *uniform* soil consists of a very narrow range of particle sizes, whereas a *well-graded* soil contains a wide range.

For a rapid estimate of grading, a *field-settling* test may be carried out in a tall jar or bottle. A sample of soil is shaken up with water in a jar and allowed to stand for a few minutes. The coarsest particles settle to the bottom first, so that a subsequent examination of the layers in the jar will yield approximate proportions of the various size ranges.

If over 65% of the soil particles are greater than 0.06 mm, the soil should be described as *coarse-grained*, i.e. either a SAND or a GRAVEL. The basis for deciding composite types is given in *Fig. 8.4*. If over 35% of the soil is less than 0.06 mm, it is *fine-grained*, i.e. SILT or CLAY.

Amount of fines. The quick-settling test will give an indication of the amount of fine-grained material present, but for a more accurate assessment a wet sieve analysis is needed (Section 8.3). To distinguish between silt and clay, rub the wet soil between the fingers: clay tends to stick to the fingers, whereas silt only leaves them slightly dusty.

In the case of composite types, it is necessary to indicate the amount of fine material in predominantly sands or gravels, or the amount of sand or gravel in silts and clays. The descriptive terms given in *Table 8.2* are used for this purpose.

Plasticity and consistency If the soil particles stick together when wet, the soil possesses **cohesion**; if the wet mass can be easily moulded it possesses **plasticity**. Grades of consistency ranging between soft and very stiff may be identified (*Table 8.3*)

Dilatancy A pat of moist soil is placed on the open palm of one hand and the side of this hand tapped with the other. Dilatancy is exhibited when, as a result of the tapping, a glossy film of water appears on the surface of the pat. When the pat is gently pressed, the water disappears and the pat becomes stiff. Very fine sands and inorganic silts exhibit marked dilatancy.

TABLE 8.4 British Soil Classification System for engineering purposes (Table 8: BS5930: 1981)

Soil groups		*Subgroups and laboratory identification*					
	GRAVEL and SAND may be qualified Sandy GRAVEL and Gravelly SAND, etc. where appropriate	*Group symbol*		*Subgroup symbol*	*Fines (% less than 0.06 mm)*	*Liquid limit %*	*Name*
COARSE SOILS less than 35% of the material is finer than 0.06 mm	Slightly silty or clayey GRAVEL	G	GW	GW	0 to 5		Well graded GRAVEL
			GP	GPu GPg			Poorly graded/Uniform/Gap graded GRAVEL
	Silty GRAVEL	G-F	G-M	GWM GPM	5 to 15		Well graded/Poorly graded silty GRAVEL
	Clayey GRAVEL		G-C	GWC GPC			Well graded/Poorly graded clayey GRAVEL
	Very silty GRAVEL	GF	GM	GML, etc	15 to 35		Very silty GRAVEL; subdivide as for GC
	Very clayey GRAVEL		GC	GCL GCI GCH GCV GCE			Very clayey GRAVEL (clay of low, intermediate, high, very high, extremely high plasticity)
	Slightly silty or clayey SAND	S	SW	SW	0 to 5		Well graded SAND
			SP	SPu SPg			Poorly graded/Uniform/Gap graded SAND
	Silty SAND	S-F	S-M	SWM SPM	5 to 15		Well graded/Poorly graded silty SAND
	Clayey SAND		S-C	SWC SPC			Well graded/Poorly graded clayey SAND
	Very silty SAND	SF	SM	SML, etc	15 to 35		Very silty SAND; subdivided as for SC
	Very clayey SAND		SC	SCL SCI SCH SCV SCE			Very clayey SAND (clay of low, intermediate, high, very high, extremely high plasticity)

FINE SOILS more than 35% of the material is finer than 0.06 mm	Gravelly or sandy SILTS and CLAYS 35% to 65% fines	Gravelly SILT	FG	MG	MLG, etc			Gravelly SILT; subdivide as for CG
		Gravelly CLAY		CG	CLG		<35	Gravelly CLAY of low plasticity
					CIG		35 to 50	of intermediate plasticity
					CHG		50 to 70	of high plasticity
					CVG		70 to 90	of very high plasticity
					CEG		>90	of extremely high plasticity
		Sandy SILT	FS	MS	MLS, etc			Sandy SILT; subdivide as for CG
		Sandy CLAY		CS	CLS, etc			Sandy CLAY; subdivide as for CG
	SILTS and CLAYS 65% to 100% fines	SILT (M-SOIL)	F	M	ML, etc			SILT; subdivide as for C
		CLAY		C	CL		<35	CLAY of low plasticity
					CI		35 to 50	of intermediate plasticity
					CH		50 to 70	of high plasticity
					CV		70 to 90	of very high plasticity
					CE		>90	of extremely high plasticity
ORGANIC SOILS		Descriptive letter 'O' suffixed to any group or sub-group symbol.			Organic matter suspected to be a significant constituent. Example MHO: Organic SILT of high plasticity.			
PEAT		Pt	Peat soils consist predominantly of plant remains which may be fibrous or amorphous.					

(Reproduced by permission of the British Standards Institution)

TABLE 8.5 Sub-group symbols in the BSCS

	Primary letter	Secondary letter
Coarse-grained soils	G = Gravel S = Sand	W = well-graded P = poorly-graded Pu = uniform Pg = gap graded
Fine-grained soils	F = Fines (Undifferentiated) M = Silt C = Clay	L = low plasticity I = intermediate plasticity H = high plasticity V = very high plasticity E = extremely high plasticity
		U = upper plasticity range*
Organic soils	Pt = Peat	O = organic (may be suffixed to any group)

*May be used as a guide when it is not possible to designate the range of liquid limit more closely, e.g. in rapid assessment of soils.

Toughness A thread of moist soil is rolled by the palm of the hand until it dries sufficiently to crumble and break just as it reaches a diameter of 3 mm. The greater th plasticity of the soil, the more easily this can be done.

Dry strength If a pat of moist soil is dried, preferably in an oven, it will shrink and harden according to the silt or clay content. Its dry strength may be estimated by attempting to break the pat with the fingers. A high dry strength indicates a clay of high plasticity, whereas a crumbly powdery dry pat indicates a silt of low plasticity.

Penetration resistance At the surface or in trial pits, a spade, a pick or a small wooder peg driven into the soil, will give an indication of compactness (see *Table 8.3*).

Soil structure From observations in trial pits or at other exposures, the main structur characteristics may be determined; such as the spacing between beds or laminations, whether or not different materials are interbedded and whether or not the soil is fissured (*Table 8.3*).

8.3 CLASSIFICATION BY PARTICLE SIZE

In the British Standard Soil Classification the sub-groups are defined primarily by particle size (*Fig 8.1*). This is acceptable for two reasons:

(1) the particle-size distribution in a soil is simply and easily ascertained in the labor and

(2) a number of engineering properties, e.g. permeability, frost susceptibility and compressibility, are related directly or indirectly to particle size.

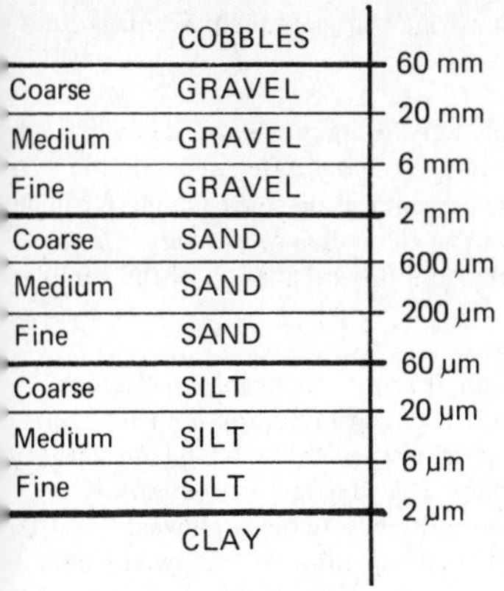

Fig 8.1 Soil groups according to particle size

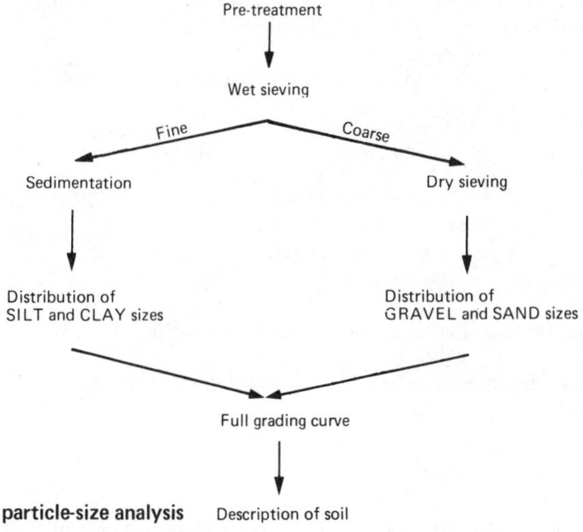

ig 8.2 Flow chart for particle-size analysis

For coarse-grained material, the particle-size distribution may be obtained by ieving, but for fine-grained (cohesive) material a sedimentation method is required. t is therefore necessary to separate the gravel/sand fractions from the silt/clay material. flow diagram of the procedure is given in *Fig 8.2*, and the various stages described riefly below. For further information and more detailed description refer to *BS 1377: lethods of testing soil for civil engineering purposes.*

re-treatment A suitably sized sample of the soil is first dried and weighed and then ieved to remove the coarsest (> 20 mm) particles. The sub-sample is then immersed n water containing a dispersing agent (e.g. a 2g/l solution of sodium hexametaphosphate).

Vet sieving After being allowed to stand, the soil/water mixture is washed through a 3 μm mesh sieve and the retained fraction again dried and weighed. The difference between the weighings before and after the wet sieving will yield the **percentage fines** for the soil. (See *Worked example 8.1*).

Dry sieving The dried retained fraction from the wet sieving procedure is now passed through a nest of sieves of mesh sizes ranging down to 63 μm. After weighing the fractions retained on each mesh, the cumulative percentages **passing** each mesh can be calculated and the coarse-grained part of the grading curve drawn. The very coarse fraction (> 20 mm) may also be sieved and included in the calculation of the cumulati[ve] percentages (see *Worked example 8.2*).

Sedimentation Because of the effects of cohesion, it is not practicable to attempt particle analysis by sieving with the fine-grained (silt or clay) fraction. A small representative sub-sample is taken from the material washed through a 63 μm sieve and made up into a suspension with distilled water. The soil/water suspension is adjusted to a volume of 500 ml and agitated vigorously before being allowed to settle in a tall jar. At specific time intervals, samples of the suspension are removed from a given depth below the surface using a special pipette. The amount of soil solid in each pipette sample is obtained by weighing.

According to Stokes' Law, the velocity of a settling particle is given by

$$\nu = \frac{d^2 (\gamma_g - \gamma_w)}{18\eta}$$

where: d = diameter of particle;
γ_g = unit weight of the particle;
γ_w = unit weight of water;
η = viscosity of water.

For average laboratory conditions: $\nu = 900\ d^2$ mm/s

Samples taken at a given depth (e.g. 100 mm) at an elapsed time of t seconds will not therefore contain particles greater than:

$$d \text{ max} \doteq \left(\frac{1}{9t}\right)^{\frac{1}{2}}$$

If this expression is used the following sampling times result:

Soil sub-group	*particle diam (d mm)*	*elapsed time*
medium sand	0.2	2.8s
find sand	0.06	31s
coarse silt	0.02	4min 38s
medium silt	0.006	51min 26s
fine silt	0.002	7hr 43min
clay		

The amount of soil in each pipette sample is expressed as a percentage and the cumulative **percentage fines** (≡ percentage passing) calculated.

Worked example 8.1 In a wet sieve analysis a specimen of soil having an initial mass of 272g was weighed after being washed on a 63 μm mesh sieve and found to have a final dried mass of 249g. Determine the percentage fines for the soil.

$$\text{Percentage fines} = \frac{\text{mass of material passing 63 }\mu\text{m mesh} \times 100}{\text{initial mass of sample}}$$

$$= \frac{272 - 249}{272} \times 100 = \mathbf{9.2\%}$$

8.4 GRADING CURVES AND CHARACTERISTICS

A *grading curve* is simply a graph of the percentage passing against particle size. Typical grading curves are shown for several different soil types in *Fig 8.3*; the percentage passing scale is linear and the size scale is logarithmic.

The primary objective is to establish a description for the soil and this may be done using the grading curve by estimating the range of sizes included in the bulk of the sample. If the sample contains 35% or more of particles less than 0.06 mm it is described as **fine-grained** (i.e. a SILT or CLAY); if 65% or more is larger than 0.06 mm it is coarse-grained (i.e. a GRAVEL or a SAND), see *Fig 8.4*.

An additional analysis is often carried out using geometric properties of the grading curve known as *grading characteristics.* First, three points (sizes) are established on the curve:

D_{10} = maximum size of the smallest 10% of the sample.
D_{30} = " " " " 30% " " "
D_{60} = " " " " 60% " " "

The grading characteristics are then defined as follows:

Effective size $= D_{10}$

Uniformity coefficient, $C_u = \dfrac{D_{60}}{D_{10}}$

Coefficient of gradation, $C_g = \dfrac{(D_{30})^2}{D_{60} \times D_{10}}$

C_u and C_g will both be unity for a single-sized soil, while $C_u < 3$ indicates **uniform** grading. Most **well-graded** soils will have grading curves that are almost flat or slightly concave, giving values of C_g between 0.5 and 2.0.

Worked example 8.2 The results of a dry-sieving test are as follows:

Sieve size (mm or μm)	14	10	6.3	5	3.35	2.0	1.18	600	425	300	212	150	63
Mass retained (g)	0	3.5	7.6	7.0	14.3	21.1	56.7	73.4	22.2	26.9	18.4	15.2	17.5

The quantity passing the 63 μm mesh sieve was 8.5g and the initial dried mass of the total sample was 292.4g. Plot the grading curve for the soil and give a classification in accordance with the BSCS.

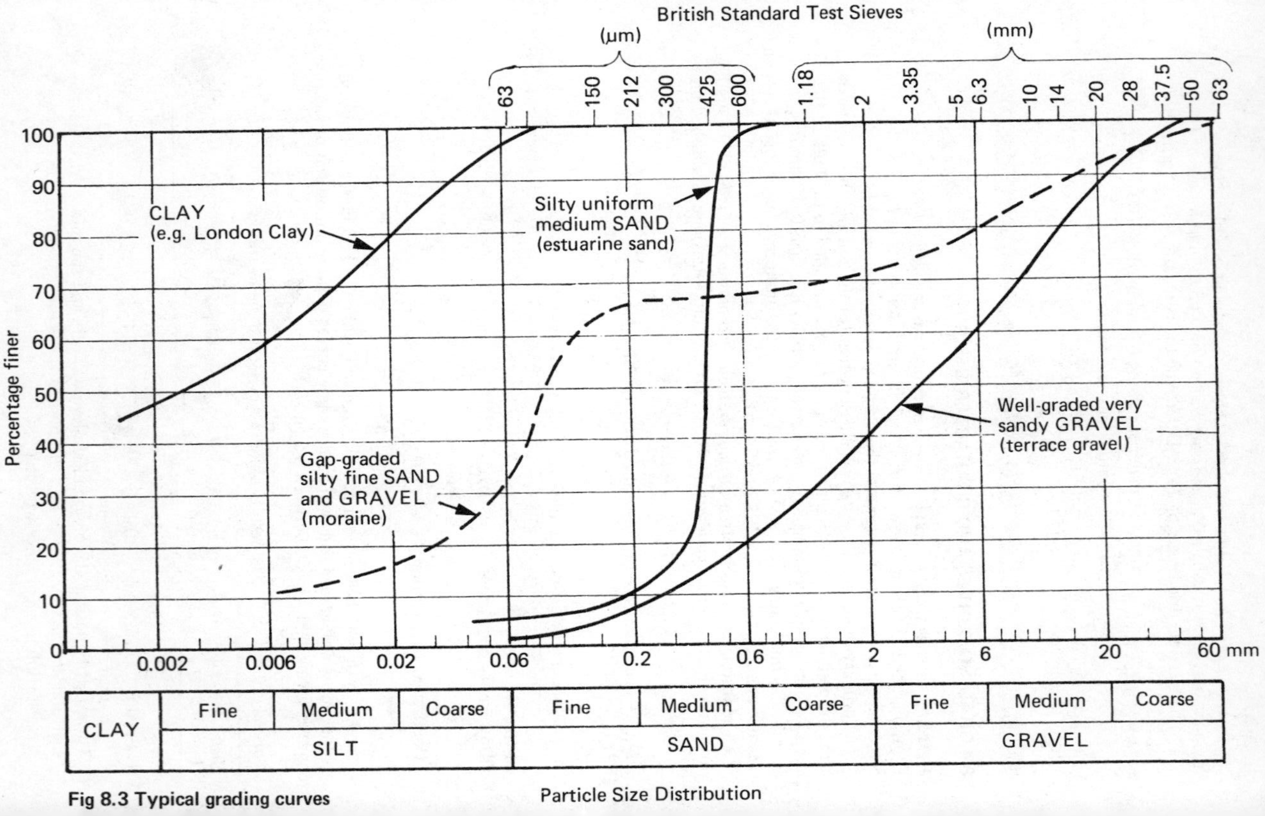

Fig 8.3 Typical grading curves

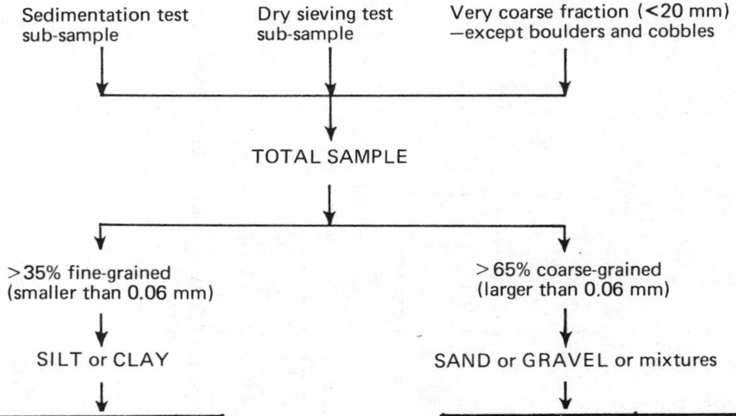

Liquid limit (%)	Plasticity term
<35	Low plasticity
35–50	Intermediate plasticity
50–70	High plasticity
70–90	Very high plasticity
>90	Extremely high plasticity

% of total sample		Description
SAND	GRAVEL	
<5		Slightly sandy GRAVEL
5–20		Sandy GRAVEL
20–30		Very sandy GRAVEL
EQUAL		GRAVEL/SAND
	20–30	Very gravelly sand
	5–20	Gravelly SAND
	<5	Slightly gravelly SAND

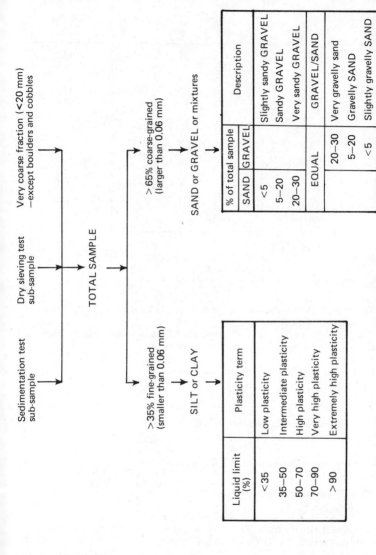

Fig 8.4 Soil description based on classification tests

The retained masses are first evaluated as percentages of the initial total mass of the sample and then the percentages passing each mesh size obtained by successive subtraction. The complete set of results is tabulated below.

	Sieve mesh size	*Mass retained* (g)	% *retained*	% *passing*
mm	14	0	0	100.0
	10	3.5	1.2	98.8
	6.3	7.6	2.6	96.2
	5.0	7.0	2.4	93.8
	3.35	14.3	4.9	88.9
	2.0	21.1	7.2	81.7
	1.18	56.7	19.4	62.3
μm	600	73.4	25.1	37.2
	425	22.2	7.6	29.6
	300	26.9	9.2	20.4
	212	18.4	6.3	14.1
	150	15.2	5.2	8.9
	63	17.5	6.0	2.9
	Pan	8.5	2.9	
Total		292.3 g	100.0%	

The total 292.3g must be checked against the initial total weight: in this case, there are no significant losses.

The plot of the grading curve is shown in *Fig 8.5*. An examination of the curve shows the soil to possess the following size fractions:

Gravel	18%	
Coarse sand	45%	79%
Medium sand	24%	
Fine sand	10%	
Silt	3%	

The **grading characteristics** obtained from the curve are:

$D_{10} = 0.163$ mm $\qquad D_{30} = 0.45$ mm $\qquad D_{60} = 1.03$ mm

$$\text{Giving Uniformity coefficient, } C_u = \frac{D_{60}}{D_{10}} = \frac{1.03}{0.163} = 6.3$$

$$\text{and Coefficient of gradation, } C_g = \frac{D_{30}{}^2}{D_{60} \times D_{10}} = \frac{0.45^2}{1.03 \times 0.163} = 1.2$$

indicating a **well-graded** soil.

In accordance with the BSCS (*Table 8.4*) the soil may be classified as a **well-graded slightly silty gravelly SAND** (Group symbol = SWG).

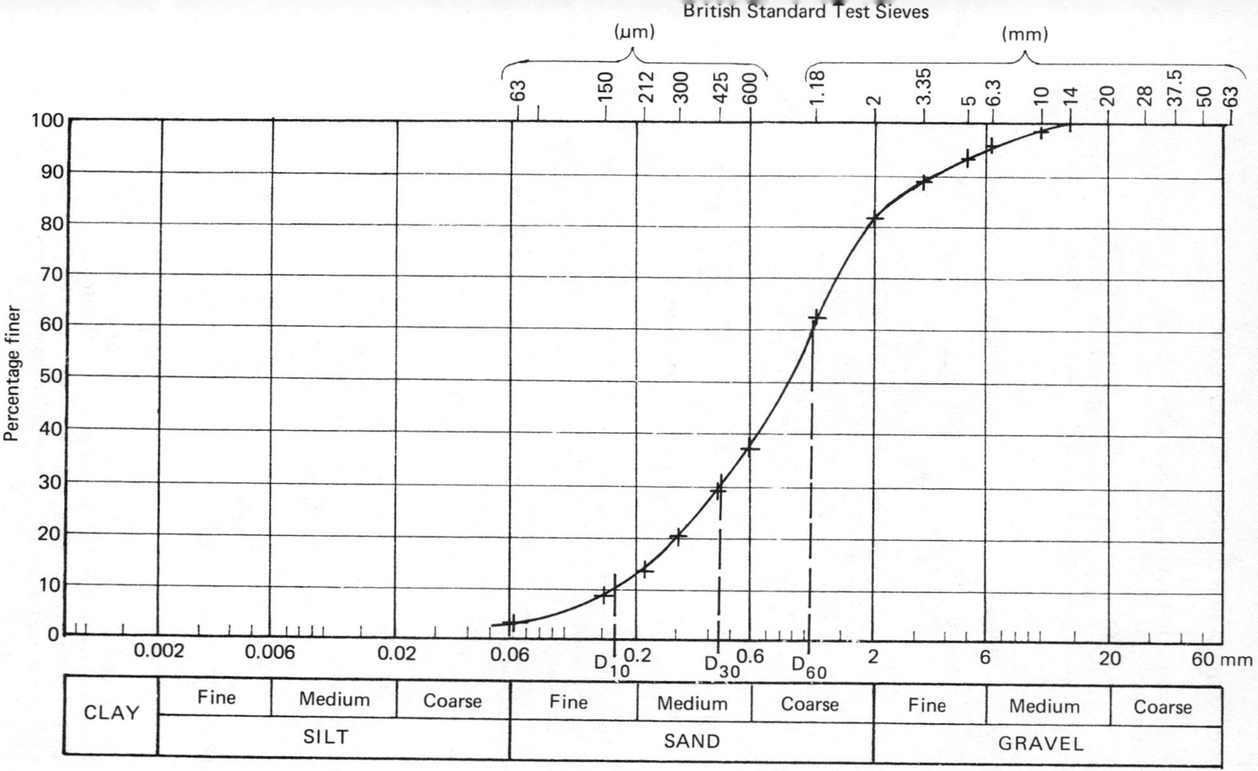

Fig 8.5

Particle Size Distribution

8.5 CLASSIFICATION OF FINE-GRAINED SOILS

The particles of fine-grained soils, particularly clays, are very flaky and so have very large surface areas. The interaction between water and flaky particles imparts the property of **plasticity** to the soil, i.e. an unrecoverable deformation results when the stress on the soil is increased. The degree of softness or stiffness in a soil depends partly on the moisture content and partly on the flakiness of the particles. (See also Chapter 9).

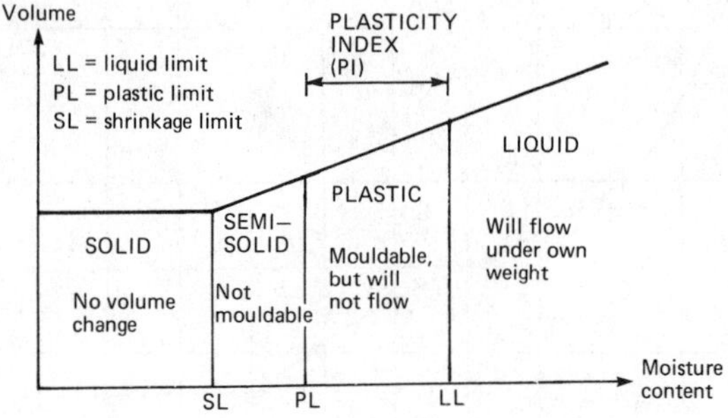

Fig 8.6 Consistency limits

Four states of consistency may be defined for cohesive soils: solid, semi-solid, plastic and liquid.

The moisture contents at which the soil changes from one consistency state to another are termed *consistency limits* (*Fig 8.6*); these form the basis of the classification.

LL = **liquid limit**, being the moisture content at which a drying soil ceases to flow as a liquid.

PL = **plastic limit**, being the moisture content at which a drying soil ceases to be plastically mouldable (i.e. begins to crumble).

SL = **shrinkage limit**, being the moisture content at which a drying soil ceases to shrink.

The upper and lower bounds of the plastic state are represented by the liquid and plastic limits respectively, and the difference between these, termed the **plasticity index**, defines the plastic range.

Plasticity index, PI = LL – PL

The sub-groups of fine-grained soil in the BSCS are defined by the relationship between the liquid limit and the plasticity index, as shown in *Fig 8.7*. The A-line on the **plasticity chart** defines an arbitary division between SILTS and CLAYS, while five degrees of plasticity are defined on the liquid limit scale (See also *Fig 8.4*).

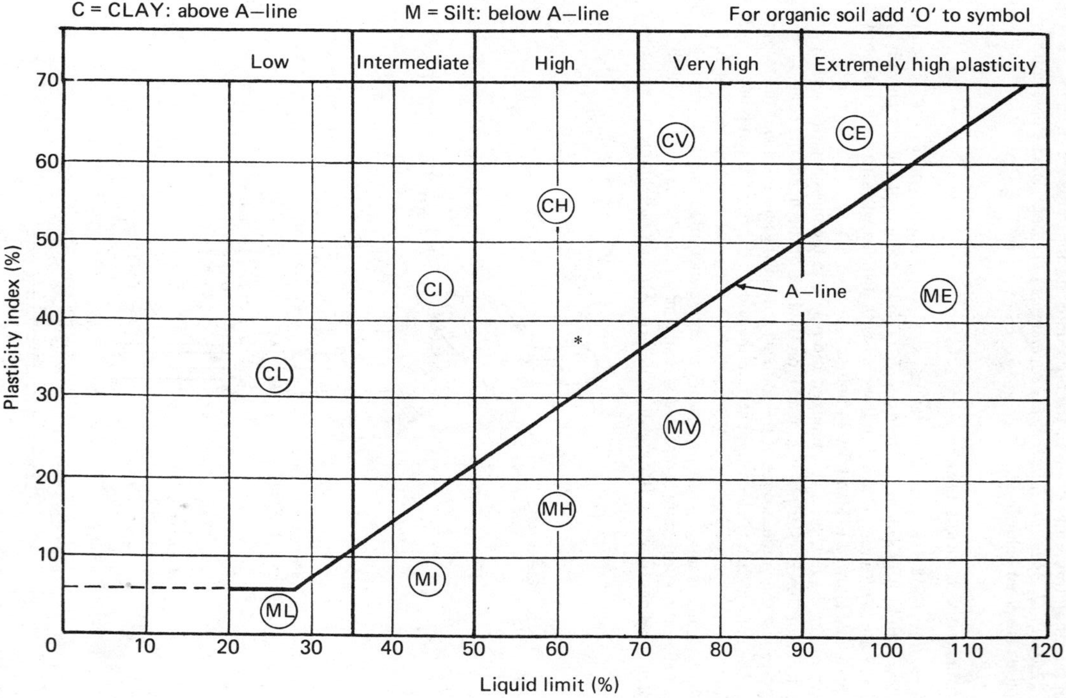

Fig 8.7 Plasticity chart for the classification of soils

8.6 DETERMINATION OF THE CONSISTENCY LIMITS

The two consistency limits used for classification may be determined by simple tests in the laboratory. Full details of the procedure and apparatus are given in *BS 1377: Methods of testing soils for civil engineering purposes*. Concise descriptions of the tests are given below.

DETERMINATION OF LIQUID LIMIT

The apparatus (*Fig 8.8*) consists of a stainless steel cone mounted on a stand, which will allow it to be dropped on to a test sample of soil. A sample of soil, which will pass a 425 μm mesh sieve, is mixed thoroughly with distilled water to form a smooth thick paste and allowed to stand for 24 hours.

To carry out the test, the soil is remixed and placed in the brass cup, the surface being struck off level. After placing the cup on the base of the stand, the cone is lowered to just touch the surface of the soil paste, the dial gauge is set and the reading noted. The cone is then released to penetrate the soil for exactly 5 seconds and relocked; a second gauge reading is now taken. The difference between the first and second dial gauge readings gives the amount of cone penetration (mm).

The penetration procedure is repeated several times on the same paste mix until consistent readings to within 0.5 mm are obtained. After a small sample of the paste

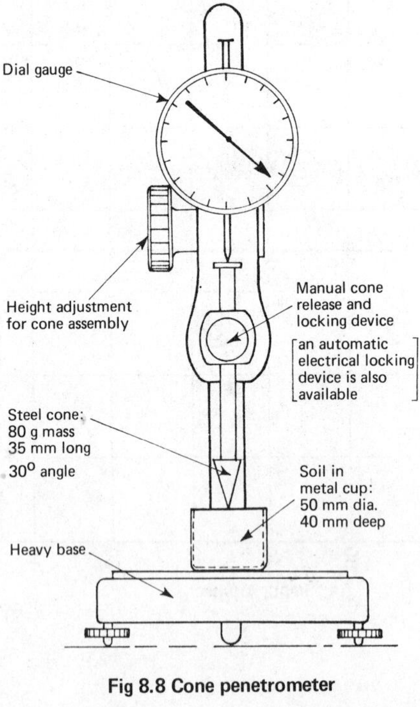

Fig 8.8 Cone penetrometer

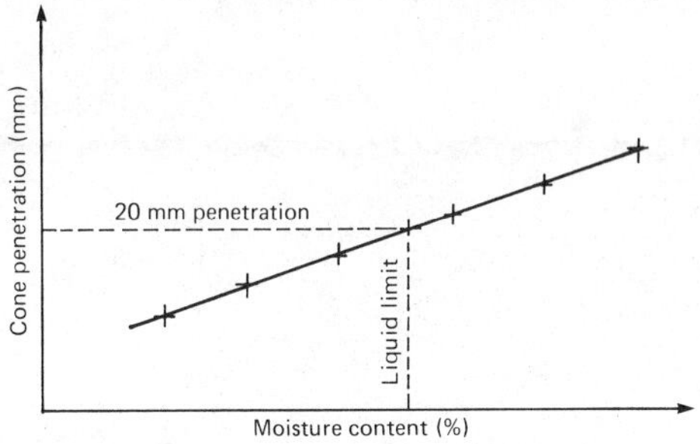

Fig 8.9 Liquid limit results plot

has been taken for moisture content determination, the whole penetration procedure is repeated another four or five times with paste mixes at different moisture contents.

The results are then plotted and a straight-line graph drawn of cone-penetration/moisture-content (*Fig 8.9*). The liquid limit of the soil is taken as the *moisture content corresponding to a penetration of 20 mm.*

An alternative method using an apparatus designed by the American, Albert Casagrande, is also described in BS 1377; however, the cone penetrometer method is preferred, since the procedure is less prone to operator error, thus giving more reliable results.

DETERMINATION OF PLASTIC LIMIT

A small quantity of the soil paste prepared for the liquid limit test is first formed into a plastic ball and then rolled into a thin thread on a glass sheet. When the diameter of the thread has been reduced to less than 3 mm, it is reformed into a ball and rolled out again. This is repeated until the thread begins to crumble just as the diameter becomes 3 mm. It is assumed that the moisture content at this point will be equal to the plastic limit of the soil.

The process is carried out on eight specimens in all and the plastic limit taken as the average of the moisture contents.

Worked example 8.3 In a liquid limit test, using a cone penetrometer, the following readings were recorded:

Cone penetration (mm)	14.4	16.4	18.2	21.1	22.3
Moisture content (%)	30.9	42.0	51.8	68.2	77.6

In a plastic limit on the same soil, the plastic limit was found to be 24%. Determine the liquid limit and the plasticity index of the soil, and suggest a classification in accordance with the British Soil Classification System.

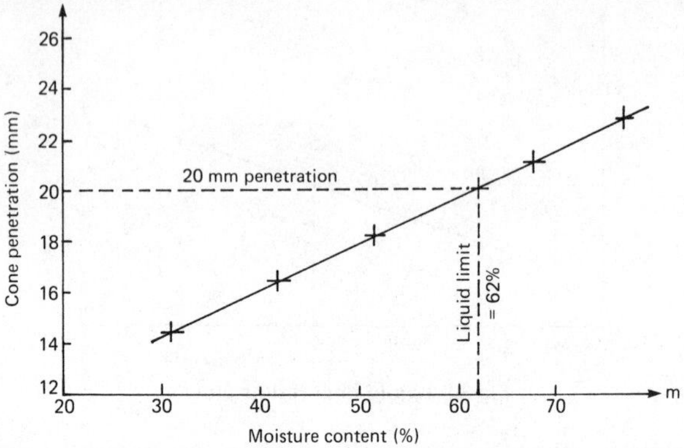

Fig 8.10

A graph of the cone penetration against moisture content is shown plotted in *Fig 8.10*. From the graph, the liquid limit is read off corresponding to a penetration of 20 mm; **LL = 62%**

Since the plastic limit was found to be 24% then

Plasticity index, = LL − PL = 62 − 24 = 38

Using the Plasticity Chart (*Fig 8.7*), the soil (indicated by an *) is seen to fall into the category of a **CLAY of high plasticity**. (Group symbol = CH).

TEST EXERCISES

COMPLETION EXERCISES (answers given on page 207)

Complete the following statements by inserting the most appropriate word or words in the spaces indicated.

1 State *three material* characteristics of a soil which may be used in its description.
(i) . . . . (ii) . . . . (iii) . . . .

2 State *three structural* characteristics of a soil which may be used in its description.
(i) . . . . (ii) . . . . (iii) . . . .

3 In the *British Soil Classification System* (BSCS) the **fine-grained** soils are finer than . . . . mm

4 In the BSCS, **coarse-grained** soils are sub-divided according to . . . . and **fine-grained** soils are sub-divided according to . . . .

5 Which sub-groups are indicated by the following symbols: CE, GWM, SCL.

6 State **five** of the characteristics which may be used in a **rapid** classification of a soil.

7 According to Stoke's Law, the settling velocity of soil particles in a water suspension is proportional to the square of their . . . .

8 The effective size of a soil is the maximum size of the smallest . . . . of the sample.

9 The uniformity coefficient of a soil is . . . . divided by the effective size.

10 A uniform soil would have a uniformity coefficient value of . . . .

11 The liquid limit of a soil is the moisture content at which it ceases . . . .

12 The plasticity index of a soil is the difference between the . . . . and . . . . limits.

13 Soils having a liquid limit in the range 50% to 70% are classified as having . . . . . . plasticity.

14 In the plasticity chart used for soil classification, the A-line separates . . . . from . . . .

ESSAY QUESTIONS

1 Discuss the aims and purposes of site investigation.

2 Describe the British Soil Classification System as set out in BS 5930:1981, and comment on where its use is particularly recommended.

3 Explain the basis for the rapid field identification and description of soils, and summarise the rapid tests used in this connection.

4 Sketch a series of grading curves to typify each of the following soil types: a silty uniform fine SAND, a well-graded gravelly SAND; a gap-graded silty gravelly medium SAND, a well-graded CLAY-SILT.

5 What are the consistency limits of a soil? How are these used to sub-divide fine-grained soils in the British Soil Classification System?

CALCULATION QUESTIONS (answers to page 207)

1 The results are given below of a dry sieving test in which the initial total sample was 38.2 g.

Sieve size (μm)	600	425	300	212	150	63
Mass retained (g)	0	0.5	1.1	1.9	5.0	23.0

The quantity passing the 63 μm mesh sieve was 6.7 g. Plot the grading curve for the soil and suggest a classification in accordance with the BSCS.

2 The result of a wet and dry sieving test are given below as percentages of the original total sample mass retained on certain mesh sizes. Plot the grading curve for the soil, establish the grading characteristics D_{10}, D_{30} and D_{60} and then suggest a classification description for the soil in accordance with the BSCS.

Sieve size (mm)	28	20	14	10	6.8	5.0	3.35	2.0	1.18
Percentage retained	0	3.7	4.9	5.3	7.7	3.4	4.7	1.5	1.6

Sieve size (μm)	600	425	300	212	150	63
Percentage retained	2.0	2.4	8.6	11.3	14.0	17.3

Percentage passing 63 μm mesh = 11.6.

3 The results of laboratory consistency tests on a particular soil are given below. Determine the liquid limit and plasticity index of the soil and suggest a classificatio in accordance with the BCCS.

Cone penetrometer results

Cone penetration (mm)	15.6	17.6	19.0	20.9	22.8
Moisture content (%)	29.6	38.8	44.6	52.5	61.1

Plastic limit test Final average plastic limit = 22%

4 Using the Plasticity Chart for the classification of fine-grained soil, determine the BSCS sub-group symbols for the following soils.
(a) LL = 62% PL = 28%
(b) LL = 78% PL = 41%
(c) LL = 47% PL = 31% Organic matter 8.2%
(d) A sand containing 17.8% clay, which has a LL of 56% and a PL of 26%.
(e) A poorly-graded gravel containing 6.2% silt, which has a LL of 38% and a PL of 19%.

9 Basic soil properties

9.1 COMPOSITION OF SOIL

The definition of the term **soil** from an engineering point of view includes those layers of loose unconsolidated material that can be worked without drilling or blasting. Geologists, agriculturalists and others may prefer alternative definitions which serve their own purposes better. Since soil consists largely of the weathered and broken-down

GAS	Air	Mass assumed to be zero. Expelled during compaction.
	Water vapour	Will freeze in contact with ice.
LIQUID	Water	Assumed to be incompressible under field conditions and to have no shear strength.
	Dissolved salts	Such as calcium or magnesium sulphate which have a deleterious effect on setting concrete.
SOLID	Granular rock/mineral particles	Rock or mineral fragments, such as quartz grains, which may be rounded or angular.
	Flaky mineral particles	Mainly clay minerals resulting from the weathering of flaky rock minerals such as mica.
	Biogenic fragments	Partially decomposed plant or animal material, or man made artifacts: common in fill and reclaimed land.
	Intergranular cement	Mostly fine-grained oxides or sulphides, insoluble or crystallised salts (soluble).
	Organic matter	Decomposed plant or animal material (humus, peat). Highly compressible, high water absorption; may affect setting of concrete.

Fig 9.1 Composition of soil

remains of surface rocks, it follows that it is made up mostly of inorganic material mixed with water and air.

It is convenient to consider a soil model which has three phases: solid, liquid and gas. In *Fig 9.1* the main constituents of the three phases are shown, together with some of their properties.

CLAY MINERALS

Clay minerals result from the weathering of feldspars and micas (see Chapter 3) and form part of a group known as **layer-lattice** minerals. The most significant property of the clay minerals is their **flakiness**. Clay flakes are both very small and extremely thin, and therefore have exceptionally large surface areas. In *Table 9.1* approximate

TABLE 9.1 Approximate surface properties of soil minerals

Mineral	*Specific surface* (m^2/g)	*Adsorbed moisture* (%)
Cubes or spheres of quartz:		
1 mm dia.	0.0022	
0.1 mm dia.	0.022	negligible
0.01 mm dia.	0.22	
0.001 mm dia.	2.2	
Sand (0.06–2.0 mm)	0.04 – 0.001	negligible
Silt (0.002–0.6 mm)	1.0 – 0.04	0.1
Clays (< 0.002 mm)		
kaolinite	20	1
illite	80	4
montmorillonite	800	40

comparative values are given for the specific surface of both rounded and flaky particles of similar size. It will be seen that, as a result of their flakiness, clays may have a surface area as much as 400 times greater than equivalent-sized rounded grains.

The surfaces of clay flakes are not only large, they also carry a **negative electrical charge**. Now water molecules are bipolar, i.e. one end is positive and the other negative, so that their positive ends are attracted and held by a negative surface. Several layers of this **adsorbed water** may be held adjacent to the clay particles; the greater the specific surface the greater the amount held. Some approximate figures are given in *Table 9.1*.

This tendency for water to be attracted to the particle surfaces gives rise to a **suction** or **negative pore pressure**. Such soils will therefore absorb water until the internal pore pressure exactly balances the suction; the moisture content at this point of balance is termed the **equilibrium moisture content (emc)**. Any change in externally applied pressure or temperature will cause a moisture content change. Under constant conditions of pressure and temperature, a clay soil having a moisture content less than its **emc** will tend to take in water and swell. Shrinkage will take place if an external compression pressure is applied or if the temperature is raised (as when drying the soil).

The swelling/shrinking capacities of clay soils depends almost entirely on the type of clay mineral present; those with high specific surfaces, such as illite and montmorillonite produce very marked swelling/shrinking characteristics. Such **shrinkable clays** may be found in the geological formations of the Lower Lias, Oxford Clay, Kimmeridge Clay, Gault Clay, London Clay and also in the Woolwich and Reading Beds.

Plasticity and cohesion are the most characteristic properties of clay soils, and these result directly from the flaky nature of the particles and the suction due to water adsorption. At low moisture contents the suction is high and most of the water is in the adsorbed layers, resulting in high cohesion and a stiff or crumbly consistency (i.e. **solid** or **semi-solid**, see Chapter 8). The soil becomes **plastic** and the cohesion falls as the moisture content increases and the particles can slide past each other. The cohesion therefore depends upon both the flakiness of the clay particles and on the moisture content at any given time.

9.2 THE THREE-PHASE MODEL AND BASIC QUANTITIES

Measures are required for a number of basic soil properties in the solution of problems, and it is convenient for this purpose to use the three-phase model. As shown in *Fig 9.2*, the model is based on a solid phase having unit volume, with the masses and volumes

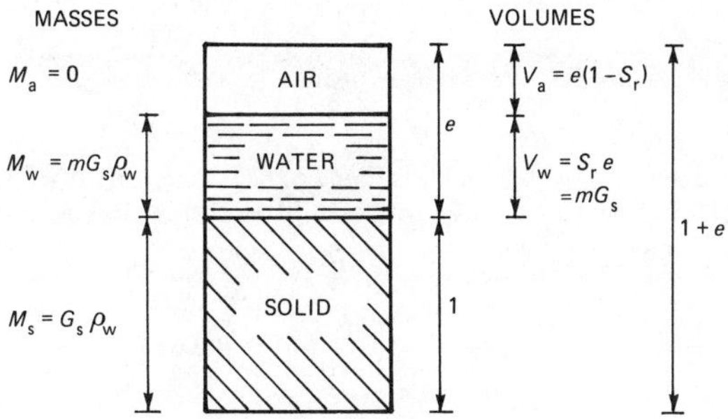

Fig 9.2 Three-phase soil model

of the other two phases given in proportion. Using this model as a basis the following volumetric quantities may be defined:

Void ratio, e $= \dfrac{\text{volume of voids*}}{\text{volume of solids}}$

(* volume of voids = volume not occupied by solids)

Porosity, n $= \dfrac{\text{volume of voids}}{\text{total volume}} = \dfrac{e}{1+e}$

Degree of saturation, S_r $= \dfrac{\text{volume of water}}{\text{volume of voids}}$

Percentage saturation $= 100\,S_r$
For a perfectly dry soil, $S_r = 0$
For a saturated soil, $S_r = 1$

Air-void ratio, $A_v = \dfrac{\text{air-void volume}}{\text{total volume}} = \dfrac{e - S_r e}{1 + e}$

$$= n(1 - S_r)$$

Also, since $S_r = \dfrac{mG_s}{e}$ (see below), $A_v = \dfrac{e - mG_s}{1 + e}$

Worked example 9.1 A soil sample is found to have a void ratio of 0.785 and a degree of saturation of 92%. Determine its porosity and air-void ratio.

Porosity, $n = \dfrac{e}{1 + e} = \dfrac{0.785}{1.785} = \mathbf{0.440}$

Air-void ratio, $A_v = n(1 - S_r)$

$$= 0.440\left(1 - \frac{92}{100}\right) = 0.035$$

(3.5%)

Worked example 9.2 A soil sample is found to have a porosity of 38%, a moisture content of 18.2% and a specific gravity of 2.70. Determine its void ratio, degree of saturation and air-void ratio.

From $n = \dfrac{e}{1 + e}$, $e = \dfrac{n}{1 - n} = \dfrac{0.38}{1 - 0.38} = \mathbf{0.613}$

Degree of saturation, $S_r = \dfrac{mG_s}{e} = \dfrac{0.182 \times 2.70}{0.613} = 0.802$

(80.2%)

Air-void ratio. $A_v = n(1 - S_r) = 0.38(1 - 0.802) = 0.075$

(7.5%)

or $A_v = \dfrac{e - mG_s}{1 + e} = \dfrac{0.613 - 0.182 \times 2.70}{1.613} = 0.075$

(7.5%)

Now if the **specific gravity** of the solids is G_s and the density of water ρ_w:

Mass of solids, $M_s = G_s\rho_w$

and Mass of water, $M_w = S_r e \rho_w$

So that the following quantities involving mass or weight may be defined:

Moisture content, $m = \dfrac{\text{mass of water}}{\text{mass of solids}}$

or $m = \dfrac{S_r e \rho_w}{G_s \rho_w}$

giving $m\,G_s = S_r e$ (a useful relationship)

Percentage moisture content = 100 m

Worked example 9.3 Determine the void ratio of a saturated soil sample which has a mass of 177.6 g before and 141.4 g after drying in an oven. Assume the specific gravity of the soil solids to be 2.68.

Moisture content, $m = \dfrac{\text{Wet mass-dry mass}}{\text{Dry mass}}$

$= \dfrac{177.6 - 141.4}{141.4} = 0.256$

Then since for a saturated soil $S_r = 1.0$

Void ratio, $e = mG_s$

$= 0.256 \times 2.68 = \mathbf{0.686}$

Dry density, $\rho_d = \dfrac{\text{mass of solids}}{\text{total volume}} = \dfrac{G_s \rho_w}{1 + e}$

Bulk density, $\rho = \dfrac{\text{total mass}}{\text{total volume}}$

$= \dfrac{\text{mass of solids + mass of water}}{\text{total volume}}$

$= \dfrac{G_s \rho_w + S_r e \rho_w}{1 + e} = \dfrac{G_s + S_r e}{1 + e} \rho_w$

The ratio of these two densities provides another useful relationship

$$\frac{\rho}{\rho_d} = \frac{\dfrac{G_s + S_r e}{1 + e} \rho_w}{\dfrac{G_s}{1 + e} \rho_w} = 1 + \frac{S_r e}{G_s}$$

or since $S_r e = mG_s$ $\rho = (1 + m)\rho_d$

Saturated density
(i.e. when $S_r = 1$) $\rho_{sat} = \dfrac{G + e}{1 + e} \rho_w$

When a volume of soil is submerged (i.e. is below the water table), it displaces an equal volume of water, so that its **effective** or **submerged density** is

$\rho' = \rho_{sat} - \rho_w$

Worked example 9.4 Determine the dry and bulk densities of a soil sample having porosity of 0.35 and a moisture content of 22%. ($G_s = 2.70$)

$$\text{Void ratio, } e = \frac{n}{1-n} = \frac{0.35}{1-0.35} = \mathbf{0.538}$$

$$\text{Dry density, } \rho_d = \frac{G_s\,\rho_w}{1+e}$$

$$= \frac{2.70 \times 1.0}{1.538} = \mathbf{1.756\ Mg/m^3}$$

$$\text{Bulk density, } \rho = \rho_d\ (1+m)$$
$$= 1.756 \times 1.22 = \mathbf{2.142\ Mg/m^3}$$

Worked example 9.5 Determine the saturated bulk density of the soil in *Worked example 9.4*, assuming no volume change.

If the soil is saturated than $S_r = 1.0$
and at constant volume, $\rho_{sat} = \dfrac{G_s + e}{1+e}\,\rho_w$

$$= \frac{2.70 + 0.538}{1.538} = \mathbf{2.105\ Mg/m^3}$$

The weight of a unit volume of soil is referred to as its **unit weight**, the units of which will be **force per unit volume** (kN/m³). The unit weights relating to appropriate densitie are given below ($g = 9.81$ m/s²).

Dry unit weight, $\gamma_d = \rho_d g$ kN/m³
Bulk unit weight, $\gamma = \rho g$ kN/m³
Saturated unit weight $\gamma_{sat} = \rho_{sat} g$ kN/m³
Unit weight of water $\gamma_w = \rho_w g = 9.81$ kN/m³
Submerged unit weight $\gamma' = \rho' g$ kN/m³

Worked example 9.6 A laboratory soil sample is found to have the following basic properties: moisture content = 22.7% grain specific gravity = 2.70. bulk density = 1.955 Mg/m³. Determine for the same sample the following other properties: dry density, void ratio, porosity, degree of saturation, air-void ratio, dry unit weight, bulk unit weight, submerged unit weight, (assuming no volume change).

Dry density, $\rho_d = \rho/(1 + m) = 1.955/(1 + 0.227) = \mathbf{1.593\ Mg/m^3}$

From $\rho_d = \dfrac{G_s\rho_w}{1+e}$

Void ratio, $e = \dfrac{G_s\rho_w}{\rho_d} - 1$

$$= \frac{2.70 \times 1.0}{1.593} - 1 = \mathbf{0.695}$$

Porosity, $n = \dfrac{e}{1+e} = \dfrac{0.695}{1.695} = 0.410\ \mathbf{(41.0\%)}$

Degree of saturation, $S_r = \dfrac{mG_s}{e} = \dfrac{0.227 \times 2.70}{0.695} = 0.882\ \mathbf{(88.2\%)}$

Air-void ratio, $A_v = \dfrac{e - mG_s}{1 + e}$

$$= \frac{0.695 - 0.227 \times 2.70}{1.695} = 0.048 \quad \mathbf{(4.8\%)}$$

Dry unit weight, $\gamma_d = \rho_d g = 1.593 \times 9.81 = \mathbf{15.63\ kN/m^3}$

Bulk unit weight, $\gamma = \rho g = 1.955 \times 9.81 = \mathbf{19.18\ kN/m^3}$

Submerged density, $\rho' = \rho_{sat} - \rho_w = \dfrac{G_s + e}{1+e}\rho_w - \rho_w$

$$= \frac{270 + 0.695 \times 1.0}{1.695} - 1.0 = \mathbf{1.003\ Mg/m^3}$$

Submerged unit weight $\gamma' = \rho' g = 1.003 \times 9.81 = \mathbf{9.84\ kN/m^3}$

9.3 DETERMINATION OF BASIC PROPERTIES

To facilitate the assessment of basic soil properties a number of laboratory tests can be carried out, some providing direct measures and others indirect measures of the property involved. Some of the more common tests are summarised below; full details may be obtained by referring to BS 1377.

MOISTURE CONTENT

A clean container (e.g. bottle, dish, tin) is first weighed empty (M_o) and then again when containing a small specimen of the wet soil (M_w). The specimen is then dried to constant weight: either in a drying oven for 24 hours at 104°C, or for a shorter period in a microwave oven. The container (with the dry soil) is now weighed again (M_d)

Moisture content, $m = \dfrac{\text{mass of water}}{\text{mass of solid}} = \dfrac{M_w - M_d}{M_d - M_o}$

Percentage moisture content $= 100\ m$

Worked example 9.7 A metal tin used in a moisture content determination had a mass of 16.56 g. When the moist soil was placed in the tin its mass was found to be 32.8 g and after oven-drying the tin had a mass of 29.48 g. Calculate the percentage moisture content of soil.

$$\text{Moisture content, } m = \frac{32.86 - 29.48}{29.48 - 16.56} = 0.262$$

Percentage moisture content = **26.2%**

SPECIFIC GRAVITY

For fine-grained soils, a 50 ml density bottle may be used, but for coarse-grained soils, a 500 ml or 1000 ml special conical topped glass jar, called a pycnometer, should be used.

After weighing the empty jar (M_o), a suitably sized specimen of dried soil is placed inside and the weight again obtained (M_1). The jar is then filled with de-aired water and the contents stirred to remove any air bubbles. The jar is topped-up with water through the small hole in the conical screw-top and then weighed again (M_2). Finally, the jar is emptied and cleaned, and then filled with de-aired water and weighed again (M_3).

Specific gravity of soil particles,

$$G_s = \frac{\text{mass of soil}}{\text{mass of water displaced by soil}}$$

$$= \frac{M_1 - M_o}{(M_3 - M_o) - (M_2 - M_1)}$$

It should be noted that the range of specific gravities of common soil particles is very narrow: between 2.64 and 2.72 in general. *Guessing* a value within this range will produce an error of no more than 3%. In order to make the laboratory test worthwhile for standard soils therefore, a high level of accuracy in the procedures and weighings is necessary, i.e. better than ± 1%.

Worked example 9.8 In a specific gravity test the following weighings were recorded:

Mass of pyconometer jar = 532 g
Mass of jar when full of clean water = 1562 g
Mass of jar containing soil only = 984 g
Mass of jar containing soil & topped up with water = 1845 g

Determine the specific gravity of the soil particles.

In the expression given above the following values may be inserted:

$M_0 = 532$ g
$M_1 = 984$ g
$M_2 = 1845$ g
$M_3 = 1562$ g

Then $G_s = \dfrac{984 - 532}{(1562 - 532) - (1845 - 984)} = \mathbf{2.67}$

POROSITY AND VOID RATIO

A suitable container (e.g. a compaction mould) is weighed (M_0) and filled with water and, if its volume (V) is not known, weighed again (M_1).

The **minimum** void ratio is found by placing the mould under water and adding the soil in three equal layers, each being thoroughly **compacted** using a vibrating rammer. After taking the mould from the water, the collar is removed and the soil struck off level: the mould is now weighed again (M_2).

$$\text{Maximum bulk density, } \rho_{(max)} = \frac{\text{mass of compacted soil and water}}{\text{volume of mould}}$$

$$= \frac{M_2 - M_0}{V} = \frac{M_2 - M_0}{M_1 - M_0}\,\rho_w$$

If the soil is saturated

$$\frac{\rho_{(max)}}{\rho_w} = \frac{\rho_{sat(max)}}{\rho_w} = \frac{G_s + e_{min}}{1 + e_{min}}$$

Then minimum void ratio,

$$e_{min} = \frac{G_s - (\rho_{sat(max)}/\rho_w)}{(\rho_{sat(max)}/\rho_w) - 1}$$

and

$$n_{min} = \frac{e_{min}}{1 + e_{min}}$$

The **maximum** void ratio is determined by repeating this procedure, except instead of compacting the soil it is poured **loosely** into the mould.

$$\text{Minimum bulk density, } \rho_{(min)} = \frac{\text{mass of loose soil and water}}{\text{volume of mould}}$$

$$= \frac{M_2 - M_0}{V} = \frac{M_2 - M_0}{M_1 - M_0}\,\rho_w$$

Thus maximum void ratio,

$$e_{max} = \frac{G_s - (\rho_{sat(min)}/\rho_w)}{(\rho_{sat(min)}/\rho_w) - 1}$$

and

$$n_{max} = \frac{e_{max}}{1 + e_{max}}$$

The **actual** void ratio (e) will lie somewhere between e_{min} and e_{max}, depending on the state of compaction of the soil. The $e_{max} - e_{min}$ range in sands and gravels can be significantly broad. A convenient measure of the degree of compaction is given by the *relative density* (D_r).

$$D_r = \frac{e_{max} - e}{e_{max} - e_{min}}$$

At maximum density, $D_r = 1$, while for a soil in its loosest state, $D_r = 0$.

Worked example 9.9 In order to determine the relative density of a soil sample the following weighings were obtained using a compaction mould having a mass of 5325 g and a volume of 948 ml.
Mass of mould filled with soil poured in loosely = 6964 g
Mass of mould filled with soil dynamically compacted = 7368 g

If the dry density of the soil in-situ is 1.65 Mg/m³ and $G_s = 2.70$, calculate the relative density of the in-situ soil.

Minimum bulk density, $\rho_{min} = \frac{6964 - 5325}{948} = 1.729\ \text{Mg/m}^3$

Maximum bulk density, $\rho_{max} = \frac{7368 - 5325}{948} = 2.155\ \text{Mg/m}^3$

Maximum void ratio, $e_{max} = \frac{2.70 - 1.729}{1.729 - 1.000} = 1.332$

Minimum void ratio, $e_{min} = \frac{2.70 - 2.155}{2.155 - 1.000} = 0.472$

In-situ void ratio, $e = \frac{2.70 \times 1.0 - 1.0}{1.65} = 0.636$

Then relative density, $D_r = \frac{1.332 - 0.636}{1.332 - 0.472} = 0.809$ (81%)

9.4 COMPACTION OF SOIL

Compaction is the compression of soil involving a reduction of air-void volume, but with no change in either solid volume or water volume. Field processes of compaction include rolling, tamping or vibrating the soil in layers of specified thickness. The objectives of compaction are (a) to reduce void ratio and permeability, (b) to increase shear strength and (c) to reduce the tendency for settlement.

The effectiveness of compaction by any process will depend on several factors: the type of soil, moisture content, maximum density possible, maximum attainable density in the field, type of plant.

The maximum attainable density is governed by the moisture content of the soil. When water is added to a dry soil adsorbed water films form around the particles providing lubrication, enabling closer packing and therefore an increase in density. As more water is added a point is reached where additional water simply pushes the grains apart, decreasing the density. The **maximum dry density** therefore occurs at an **optimum moisture content.**

Tests are described in BS 1377 whereby the maximum dry density and optimum moisture content of a soil may be determined. Briefly, the procedure consists of placing the soil in a mould in either three or five equal layers, each of which is rammed in a prescribed manner.

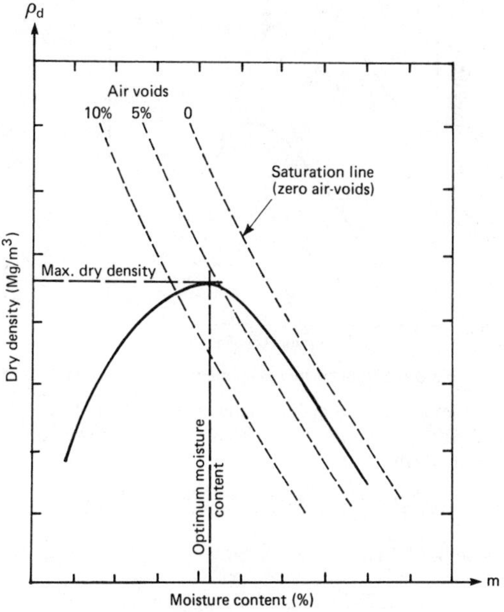

Fig 9.3 Dry density/moisture content graph

After compacting the topmost layer, the bulk density of the soil in the mould is determined and a sample taken to find its moisture content. The test is carried out on at least five samples of the same soil at different moisture contents.

From the values of bulk density and moisture content, the dry density of each compacted sample is obtained from $\rho_d = \rho/(1 + m)$ and a graph of ρ_d/m drawn as shown in *Fig 9.3.* The values of **maximum dry density** and **optimum moisture content** are then obtained from the graph.

The maximum possible dry density at a given moisture content is known as the **saturation dry density**, when the soil will have zero air voids ($A_v = 0$). For values of $A_v > 0$, the maximum attainable dry density is given by

$$\rho_d = \frac{G_s \rho_w}{1 + mG_s} (1 - A_v)$$

The air-void content may be estimated from the test results by drawing on the graph a set of curves representing the dry density at 0, 5% and 10% air voids. It is

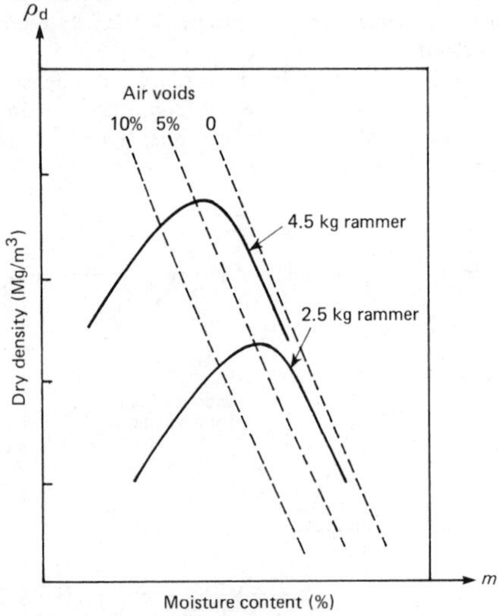

Fig 9.4 Effect of increase in compactive effort

important to note that an increase in compactive effort will produce a higher maximum dry density at a lower optimum moisture content, but with the air-void ratio remaining approximately constant (*Fig 9.4*).

Worked example 9.10 In a BS compaction test the following results were recorded:

Moisture content (%)	5.8	8.6	11.5	14.2	16.6	18.7
Bulk density (Mg/m^3)	1.867	1.996	2.100	2.140	2.094	2.108

$G_s = 2.70$

(a) Draw the graph of dry density against moisture content and from it determine the maximum dry density and optimum moisture content.

(b) On the same axes, draw the ρ_d/m curves for zero and 5% air voids, and hence determine the air-void percentage corresponding to the maximum dry density.

(a) The dry densities are obtained from $\rho_d = \rho_{1-m}$

Moisture content (%)	5.8	8.6	11.5	14.2	16.6	18.7
Dry density (Mg/m^3)	1.765	1.830	1.883	1.874	1.796	1.700

These figures are shown plotted as a graph of ρ_d/m in *Fig 9.5*.

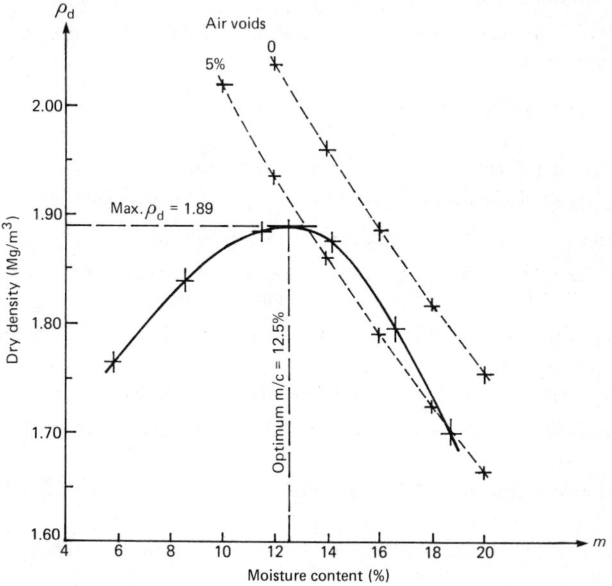

Fig 9.5

From the curve, maximum dry density = **1.89 Mg/m^3**
optimum moisture content = **12.5%**

(b) The dry densities corresponding to zero and 5% air voids are obtained from

$$\rho_d = \frac{G_s \rho_w}{1 + mG_s}(1 - A_v)$$

Moisture content (%)	10	12	14	16	18	20
ρ_d when: $A_v = 0$	2.126	2.039	1.959	1.885	1.817	1.753
$A_v = 5\%$	2.020	1.937	1.861	1.791	1.726	1.665

These two sets of figures are also plotted in *Fig 9.5*, from which, at the point of maximum dry density on the main curve, A = **6.4%**

This may be checked by calculation using the equation given above:

$$A_{v(\max \rho d)} = 1.0 - \frac{1.89\,(1 + 0.125 \times 2.70)}{2.70 \times 1.0} = 0.064\ \mathbf{(6.4\%)}$$

TEST EXERCISES

COMPLETION QUESTIONS (answers on page 207)

Complete the following statements by inserting the most appropriate word or words in the spaces indicated.

1 The three phases of the soil model are: . . . . , ... and . . . .

2 The most significant property of the clay minerals is their . . . .

3 The surfaces of clay particles carry . . . . electrical charges.

4 Adsorbed water is held on to clay particle surfaces by . . . .

5 When there is no tendency for a soil to either take up more water or to expel water it is said to be at its . . . . moisture content.

6 The two primary characteristic properties of clay soils are . . . . and . . . .

7 The void ratio of a soil is defined as the volume of voids divided by the . . . .

8 The moisture content of a soil is defined as the mass of water divided by the . . . .

9 The degree of saturation of a soil is defined as the volume of water divided by the . . . .

10 The dry density of a soil is defined as the . . . . divided by the total volume.

11 The bulk density of a soil is defined as the . . . . divided by the total volume.

12 The bulk density of a soil is given by the following expression:

$$\rho = \frac{\cdots\cdots\cdots\cdots}{\cdots\cdots\cdots\cdots}$$

13 The submerged or effective density of a soil is given by the following expression: $\rho' = \rho_{sat} - \cdots\cdots\cdots\cdots$

14 The bulk density, dry density and moisture content of a soil are related as follows: $\rho = \rho_d\,(\cdots\cdots\cdots\cdots)$.

15 The specific gravity of most soils lies between . . . . and . . . .

16 The bulk density of most soils lies between . . . . Mg/m^3 and . . . . Mg/m^3.

17 Relative density indicates the degree of . . . . of a soil.

18 In the BS Compaction Test, the primary objective is to determine for the soil the . . . . and . . . .

19 The process of compaction involves reducing the . . . . volume by rolling, tamping or vibrating the soil.

20 The **saturation dry density** is attained when the . . . . is zero.

ESSAY QUESTIONS

1 Explain the flaky nature of the clay minerals and describe how this controls the engineering properties of clay soils.

2 What are **shrinkable clays**? Where might they be found in Great Britain and what significance do they have in foundation design and construction?

3 Explain the process of compaction, commenting on the objectives, the methods used in the field and the methods of measuring the effectiveness of compaction.

CALCULATION EXERCISES (answers on page 208)

1 A soil sample is found to have a moisture content of 18%, a porosity of 39% and a grain specific gravity of 2.68. Determine the void ratio, degree of saturation and air void ratio.

2 A mass of moist soil was formed in a cylindrical mould which had a volume of 958 ml and a mass (empty) of 5248 g. The mass of the full mould was 7201 g and the moisture content of the soil was 12.8%. Assuming $G_s = 2.70$, calculate: (a) the bulk and dry densities, (b) the void ratio and porosity and (c) the degree of saturation and air void ratio.

3 A dry clean sand has a maximum density of 1.68 Mg/m^3 and a grain specific gravity of 2.68. Calculate (assuming constant volume) its bulk density and moisture content when (a) saturated (b) 50% saturated.

4 Determine the void ratio of a saturated soil sample having a moisture content of 21.6%. Assume $G_s = 2.70$.

5 A sandy soil has a porosity of 36% and a grain specific gravity of 2.67. Determine its void ratio, dry unit weight and saturated unit weight (assuming no volume change).

6 A soil sample has a porosity of 40% and bulk unit weight of 19.5 kN/m^3. Assuming $G_s = 2.68$, calculate: (a) the void ratio, (b) moisture content and degree of saturation, (c) air-void ratio and (d) dry unit weight and submerged unit weight (assuming no volume change).

7 Determine the relative density of a site soil from the data given below:

In-situ dry density	= 1.72 Mg/m^3
Minimum saturated density	= 1.78 Mg/m^3
Maximum saturated density	= 2.18 Mg/m^3
Grain specific gravity	= 2.69

8 The following data was recorded in a BS compaction test.

Moisture content (%)	9.8	11.1	12.4	13.7	15.6	17.2
Bulk density (Mg/m^3)	1.963	2.069	2.145	2.164	2.133	2.101

$G_s = 2.68$

(a) Draw the graph of dry density against moisture content and from this determine the maximum dry density and optimum moisture content.

(b) On the same graph axes, draw the curves for the dry density corresponding to zero and 5% air-voids and, hence, determine the air-void percentage in the soil at its maximum dry density.

10 Groundwater movement

10.1 GROUNDWATER FLOW AND PERMEABILITY

Since soils are loose and porous materials water is able to flow from zones of high pressure to zones of low pressure. The **permeability** of a soil is its capacity to allow water to flow through it. A measure of soil's permeability will be required when dealing with problems of seepage through or under dams, land drainage and water supply.

In saturated soil masses, flow is governed by **Darcy's Law**, which states that **the velocity of flow in a given direction is proportional to the hydraulic gradient in that direction**, i.e.

$$v = ki$$

where v = velocity of flow,
k = coefficient of permeability,
i = hydraulic gradient, $= \dfrac{\Delta H}{\Delta L}$
ΔH = change in pressure head over a flowpath length of ΔL

The quantity (Q) of water flowing is therefore

$$Q = Akit$$

or $q = Aki$

where q = quantity flowing in unit time
A = cross sectional area of flow path
t = elapsed time

The range of values of coefficient of permeability (k) is very large, extending from 1000 m/s for coarse gravels to almost nothing for clays. Also, in clay soils the presence of fissures will markedly increase overall permeability. *Table 10.1* shows the range of values of k, together with associated dranage conditions.

10.2 LABORATORY DETERMINATION OF *k*

CONSTANT HEAD TEST

This test is suitable for coarse-grained soils, such as gravels and sands, where $k > 10^{-4}$ m/s, and is carried out in a constant-head **permeameter** (*Fig 10.1*).

The soil sample is contained in a perspex cylinder with a wire-mesh disc and a layer of gravel filter above and below. Manometers are connected through the side of the cylinder so that pressure heads along the flow path may be measured (although only

TABLE 10.1 Coefficient of permeability; range of values

COEFFICIENT OF PERMEABILITY, k (m/s)											
10^{2}	10^{1}	1	10^{-1}	10^{-2}	10^{-3}	10^{-4}	10^{-5}	10^{-6}	10^{-7}	10^{-8}	10^{-9}
	CLEAN GRAVELS			CLEAN SANDS			VERY FINE SANDS			CLAY-SILTS (> 20% CLAY)	
				GRAVEL-SAND MIXTURES			SILTS AND SILTY SANDS				
				FISSURED AND WEATHERED CLAYS						UNFISSURED CLAYS	
	VERY GOOD DRAINAGE				GOOD DRAINAGE			POOR DRAINAGE			PRACTICALLY IMPERVIOUS

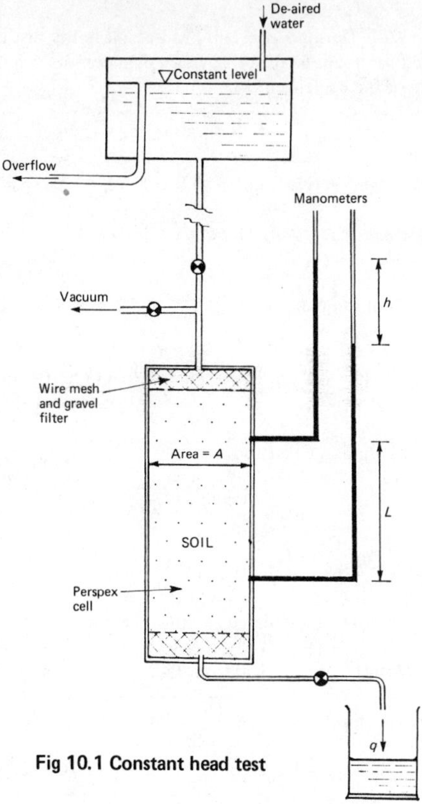

Fig 10.1 Constant head test

one pair is shown, several may be provided). Water is allowed to flow through the sample from a tank designed to maintain a constant head throughout the test. The quantity of water (Q) passing through the sample in time (t) is measured by weighing the collecting vessel.

The difference in manometer levels (ΔH) over a flowpath length (ΔL) gives the hydraulic gradient (i).

Then applying Darcy's Law: $Q = Akit$

or $$k = \frac{Q}{Ait}$$

where A = cross-sectional area of the sample.

The presence of air-bubbles can produce large errors, so that de-aired water must be used. In addition, a vacuum is often applied prior to commencing to test to remove entrapped air. Several tests at different rates of flow are carried out and the average k value reported.

Worked example 10.1 During a constant head permeability test the following data was recorded for a sample of soil having a diameter of 75 mm. Determine the average value of the coefficient k

Flow quantity for 2 min (ml)	1009	942	757	606	474
Difference in manometer levels (mm)	82	75	62	50	38

Distance between manometer tapping points = 120 mm
Length of flow path = 120 mm

Cross-sectional area of sample, $A = \frac{75^2 \times \pi}{4} = 4418 \text{ mm}^2$

Flow quantity $Q = Q(\text{ml}) \times 10^3 \text{ mm}^3$
Flow time $t = 2 \times 60$ seconds

Then applying the equation given above

$$k = \frac{Q\Delta L}{A\Delta h t} \qquad (\text{since } i = \frac{\Delta h}{\Delta L})$$

$$= \frac{Q \times 10^2 \times 120}{4418\Delta h \times 120} = 0.226 \frac{Q}{\Delta h} \text{ mm/s}$$

The calculations can now be tabulated as shown below.

Flow quantity, Q (ml)	1009	942	757	606	474
Head difference, Δh (mm)	82	75	62	50	38
k (mm/s)	2.78	2.84	2.76	2.74	2.82

Average k = 2.79 mm/s = **2.79 × 10⁻³ m/s**

FALLING HEAD TEST

The falling head test is used for fine-grained soils and soils containing a significant proportion of silt or clay, the flow rate being considerably slower than in the constant-head test.

An undisturbed sample, or a prepared compacted sample of soil in a tube of 100 mm diameter (usually), is set up as shown in *Fig 10.2*. A wire-mesh disc and a gravel filter is provided at the top and bottom. The base of the cylinder stands in a trough of water and a glass standpipe is connected to the top.

Having removed air from the apparatus, the de-aired water in the standpipe is allowed to flow through the sample and the height in the standpipe recorded at several time intervals. The test is repeated using standpipes of different diameters and an average value of k obtained.

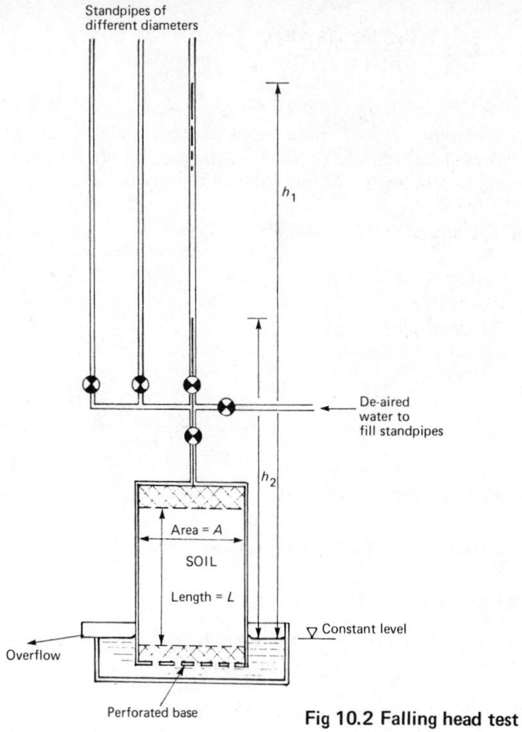

Fig 10.2 Falling head test

If a = cross-sectional area of standpipe;
A = cross-sectional area of sample;
L = length of sample.
Then quantity of water flowing in time $dt = Q = -adh$

Also from Darcy's Law, $Q = kAidt$
in which $i = h/L$

Then $$Q = -adh = \frac{kAhdt}{L}$$

Rearranging and integrating:

$$-\int_{h_1}^{h_2} \frac{dh}{h} = \frac{kA}{aL} \int_{t_1}^{t_2} dt$$

$$\therefore \quad -\log_e \frac{h_2}{h_1} = \frac{kA}{aL}(t_2 - t_1)$$

Giving $$\boldsymbol{k = \frac{aL \log_e(h_1/h_2)}{A(t_2 - t_1)}}$$

or

$$k = \frac{2.3\,aL\,\log_{10}(h_1/h_2)}{A(t_2 - t_1)}$$

Worked example 10.2 During a falling head permeability test the following data was recorded for a sample of soil having a diameter of 100 mm and a length of 150 mm. Determine the average value of the coefficient k.

Standpipe diameter = 9.00 mm

Initial standpipe level (mm)	h_1	1200	900	750
Final standpipe level (mm)	h_2	900	750	500
Time interval (s)	$t_2 - t_1$	65	41	95

Cross-sectional area of sample, A $= 100^2 \times \frac{\pi}{4}$

Cross-sectional area of standpipe, $= 9^2 \times \frac{\pi}{4}$

Using the equation given above

$$k = \frac{a\,L\,\log_e (h_1/h_2)}{A(t_2 - t_1)} = \frac{9^2 \times 150 \log_e (h_1/h_2)}{100^2 (t_2 - t_1)}$$

$$= \frac{1.215 \log_e (h_1/h_2)}{(t_2 - t_1)}$$

The three values of k obtained are:

$$k_1 = \frac{1.215 \log_e (1200/900)}{65} = 5.38 \times 10^{-3}\ \text{mm/s}$$

$$k_2 = \frac{1.215 \log_e (900/750)}{41} = 5.40 \times 10^{-3}\ \text{mm/s}$$

$$k_3 = \frac{1.215 \log_e (750/500)}{95} = 5.19 \times 10^{-3}\ \text{mm/s}$$

Average $k = 5.32 \times 10^{-3}$ mm/s $=$ **5.32×10^{-6} m/s**

10.3 DETERMINATION OF k BY PUMPING TEST

In-situ measurements of k, although more expensive and difficult to carry out, often provide more reliable results, especially in stratified or fissured deposits. The flow conditions in the aquifer are assumed to be either **unconfined** or **confined**.

An arrangement such as that shown in *Fig 10.3* is used, with the water being extracted from the pumping well at a constant measured rate (q) after steady-state conditions

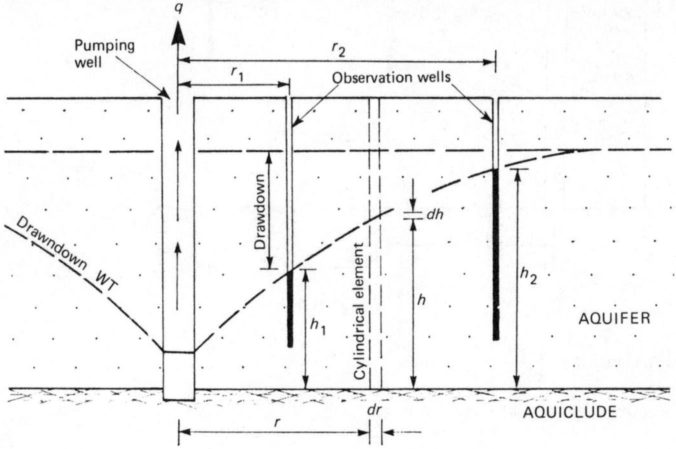

Fig 10.3 Unconfined pumping test

are obtained in the observation wells. It is assumed that the original piezometric surface was horizontal and that the hydraulic gradient at any given radius is a constant value,

$$i = \frac{dh}{dr}$$

The area through which flow occurs, $A = 2\pi r h$

Then applying Darcy's Law: $q = A\,k\,i = 2\pi r h k \dfrac{dh}{dr}$

or

$$\frac{dr}{r} = \frac{2\pi}{q} k\,h\,dh$$

Integrating

$$\int_{r_1}^{r_2} \frac{dr}{r} = \frac{2\pi k}{q} \int_{h_1}^{h_2} h dh$$

$$\therefore \quad \log_e \frac{r_2}{r_1} = \frac{2\pi k}{q} (h_2^2 - h_1^3)$$

Giving

$$\boldsymbol{k} = \frac{\boldsymbol{q}}{\pi} \frac{\log_e(r_2/r_1)}{h_2^2 - h_1^2}$$

Worked example 10.3 A pumping test was carried out to determine the permeability of a sand layer in an unconfined aquifer with a well arrangement as shown in *Fig 10.4*. At a steady-state pumping rate of 58.7m^3/h, the drawdowns in the observation wells were respectively 2.91 m and 0.88 m. Calculate the coefficient of permeability *k*.

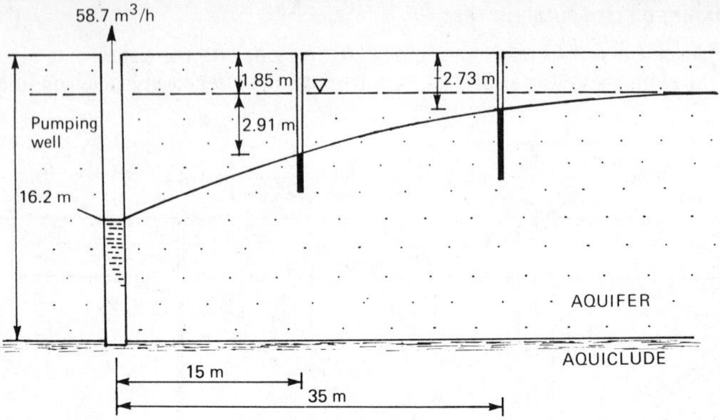

Fig 10.4

Flow rate $q = \dfrac{58.7}{60 \times 60}$ m^3/s

Reduced levels in observation wells:

$$h_1 = 16.2 - 1.85 - 2.91 = 11.44 \text{ m}$$
$$h_2 = 16.2 - 1.85 - 0.88 = 13.47 \text{ m}$$

Using the expression given above

$$k = \frac{q \log_e(r_2/r_1)}{h_2^2 - h_1^1}$$

$$= \frac{58.7 \log_e (35/15)}{3600\, \pi\, (13.47^2 - 11.44^2)} = 8.70 \times 10^{-5} \text{ m/s}$$

CONFINED-FLOW PUMPING TEST

In an aquifer confined above and below by impermeable strata, the test conditions are as shown in *Fig 10.5*, providing the piezometric surface is above the upper surface of the aquifer at all radii, both before and during the test.

Pumping is maintained at a constant rate (q) until a steady state has been achieved

Then at a given radius r, the hydraulic gradient, $i = \dfrac{dh}{dr}$

The area through which flow occurs, $A = 2\pi r D k$

Then applying Darcy's Law: $q = Aki$

$$= 2\pi r D k \frac{dh}{dr}$$

or

$$\frac{dr}{r} = \frac{2\pi k}{q} D dh$$

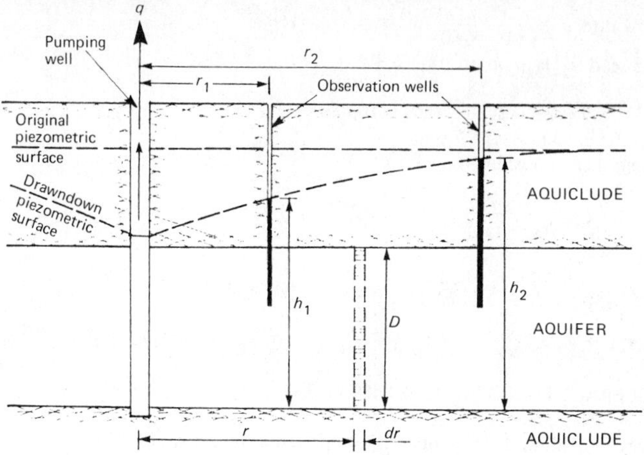

Fig 10.5 Confined pumping test

Integrating
$$\log_e \frac{r_2}{r_1} = \frac{2\pi k}{q} D(h_2 - h_1)$$

Giving
$$k = \frac{q}{2\pi D} \frac{\log_e(t_2/r_1)}{h_2 - h_1}$$

Worked example 10.4 A permeability pumping test was carried out in a confined aquifer. The arrangement of wells and all relevant dimensions is shown in *Fig 10.6*. The drawdowns indicated were observed at a steady-state pumping rate of 15.6 m^3/h.

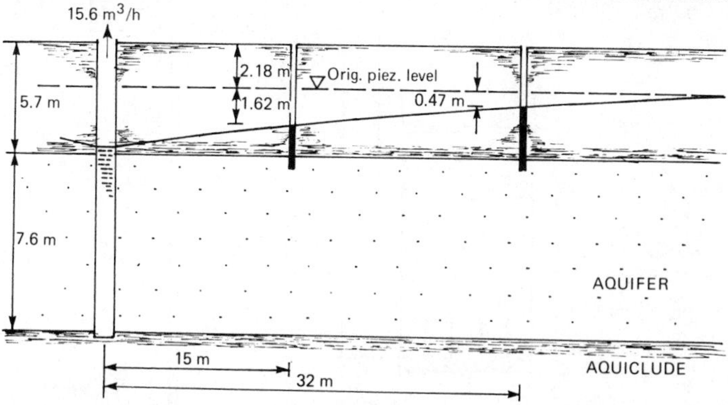

Fig 10.6

$$\text{Flow rate, } q = \frac{15.6}{60 \times 60} \text{ m}^3/\text{s}$$

Reduced levels in observation wells:

$h_1 = 5.7 + 7.6 - 2.18 - 1.62 = 9.50$ m

$h_2 = 5.7 + 7.6 - 2.18 - 0.47 = 10.65$ m

Aquifer thickness, D = 7.6 m

Using the expression given above

$$k = \frac{q}{2\pi D} \frac{\log_e (r_2/r_1)}{h_2 - h_1}$$

$$= \frac{15.6 \times \log_e (32/15)}{3600 \times 2\pi \times 7.6\,(10.65 - 9.50)} = 6.0 \times 10^{-5} \text{ m/s}$$

10.4 SEEPAGE PRESSURE AND INSTABILITY

In Darcy's equation, the velocity v is interpreted as the **apparent velocity**, i.e. velocity relative to the total cross-sectional area of flow (A). The actual velocity through the pores of the soil will be greater and is termed the **seepage velocity** (v_s).

For a soil of porosity, n, the pore area $= nA$

So, for a given rate of flow, $q = Av = nAv_s$

Hence $v_s = \frac{v}{n} = \frac{ki}{n}$

If the pressure head causing flow at a given point is h_s, then the **seepage pressure** at that point in the direction of flow will be $\gamma_w h_s$. In the case of downward seepage the effective pressure on the soil is increased, so that compression (settlement) might result. When the direction of seepage flow is upward the effective weight of the soil is decreased.

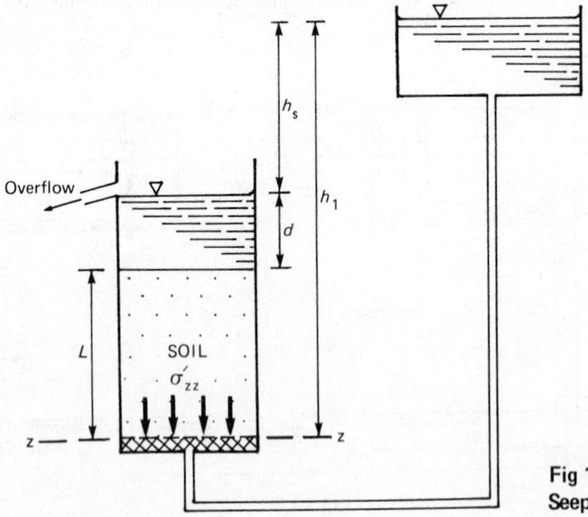

Fig 10.7
Seepage model

Consider the seepage model shown in *Fig 10.7*. The vertical effective stress at level zz is given by

$$\sigma'_{zz} = \gamma_{sat}L - \gamma_w(L + d) - \gamma_w h_1$$
$$= (\gamma_{sat} - \gamma_w)L - \gamma_w(h_1 - L - d)$$
$$= \gamma' L - \gamma_w h_s$$

It will be seen that if h_s is increased a condition can be reached where $\sigma'_{zz} = 0$, resulting in a total loss of shear strength in the case of a granular soil. This condition of instability is often referred to as the **quick** or **quicksand** condition, and the hydraulic gradient at which it occurs is called the **critical hydraulic gradient** (i_c)

When $\sigma'_{zz} = 0$, $\gamma' L = \gamma_w h_s$

Then critical hydraulic gradient, $i_c = \dfrac{h_s}{L} = \dfrac{\gamma'}{\gamma_w}$

In terms of basic soil properties:

$$i_c = \frac{\gamma'}{\gamma_w} = \frac{\gamma_{sat} - \gamma_w}{\gamma_w}$$
$$= \frac{\dfrac{G_s + e}{1 + e}\gamma_w - \gamma_w}{\gamma_w} = \frac{G_s - 1}{1 + e}$$

Worked example 10.5 It is proposed to excavate a medium-fine sand in a sheeted trench to a depth of 4.0 m, as illustrated in *Fig 10.8*. If the sand has a porosity of 38% and a grain specific gravity of 2.68, determine the maximum pumping depth required for (a) just-stable condition (b) a factor of safety of 1.25 against piping.

Void ratio, $e = \dfrac{0.38}{1 - 0.38} = 0.613$

Critical hydraulic gradient for the soil, $i_c = \dfrac{G_s - 1}{1 + e}$

$$= \frac{2.68 - 1}{1 + 0.613} = 1.042$$

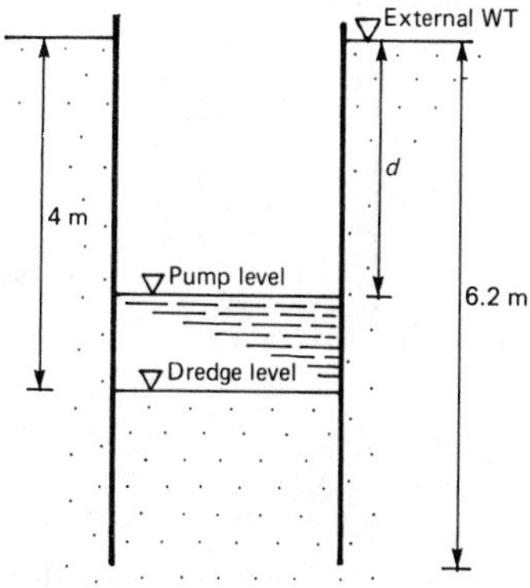

Fig 10.8

In the problem, $i = \dfrac{d}{6.2 - 4.0} = \dfrac{d}{2.2}$

(a) Then for FS = 1.0, d = 2.2 × 1.042 = **2.25 m**
(b) and for FS = 1.25, d = 2.25/1.25 = **1.80 m**

10.5 TWO-DIMENSIONAL SEEPAGE FLOW NETS

Two-dimensional flow occurs around sheet piling, under dams and through embankments the component vertical and horizontal velocities varying from point to point on the cross-section. If the soil is assumed to be homogeneous and to have the same coefficient of permeability in both the horizontal and vertical directions, a mathematical expression can be derived for the two-dimensional seepage conditions. When represented graphically, this expression takes the form of two sets of orthogonal curves, termed a **flow net.**

A flow net may be constructed by solving the equation for a large number of points within the problem, or by using an electrical analogue, or by building a model. The quickest method, however, consists of sketching the flow net in accordance with certain rules.

RULES FOR THE CONSTRUCTION OF FLOW NETS

1. **Flow lines** are drawn parallel to the direction of flow; they may *not* cross each other; the intervals between flow lines represent equal fractions (Δq) of the total flow quantity (Q).

2 **Equipotential lines** are drawn at right angles to the direction of flow and therefore must intersect the flow lines at 90°. They may not cross each other; the intervals between equipotential lines represent equal fractions (Δh) of the total head lost (h).

3 Impermeable boundaries defined in the problem are considered to be flow lines; permeable boundaries are considered to be equipotential lines.

4 The fields formed by two adjacent flow lines and two adjacent equipotential lines must be square.

Both the quantity of seepage flow and the distribution of seepage pressure can be determined from a flow net once it has been drawn.

Consider the flow net field shown in *Fig 10.9*.

Flow quantity through the field, $\Delta q = Aki$
Assuming a field interval dimension of b,

$A = b \times 1$ and $i = \dfrac{\Delta h}{b}$

The $\Delta q = b \times k \times \dfrac{\Delta h}{b} = k\Delta h$

Now total quantity flowing, $Q = N_f\, \Delta q$
and total head lost, $h = N_e\, \Delta h$

where N_f = number of flow channels
and N_e = number of equipotential drops

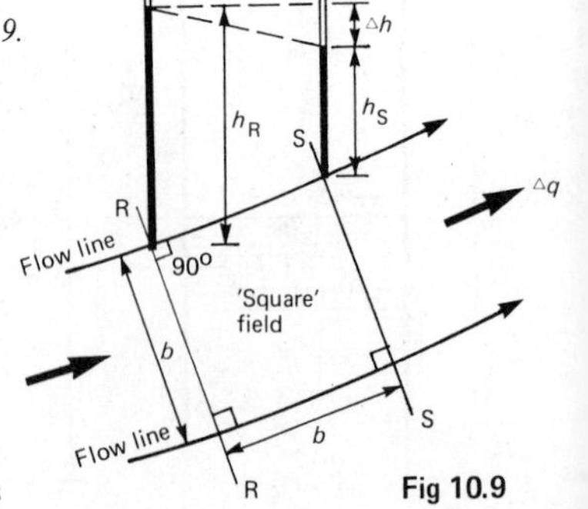

Fig 10.9

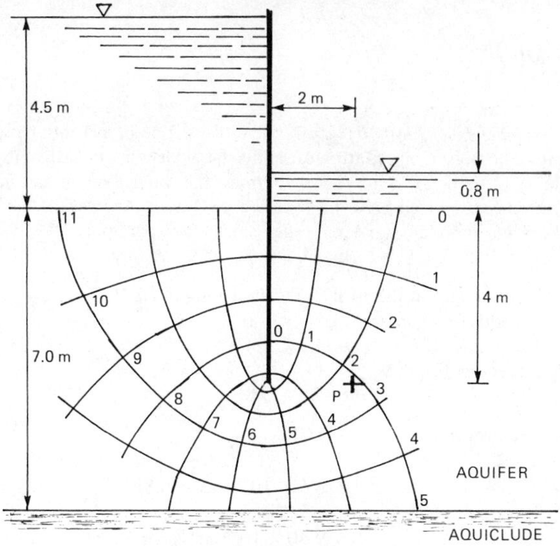

Fig. 10.10

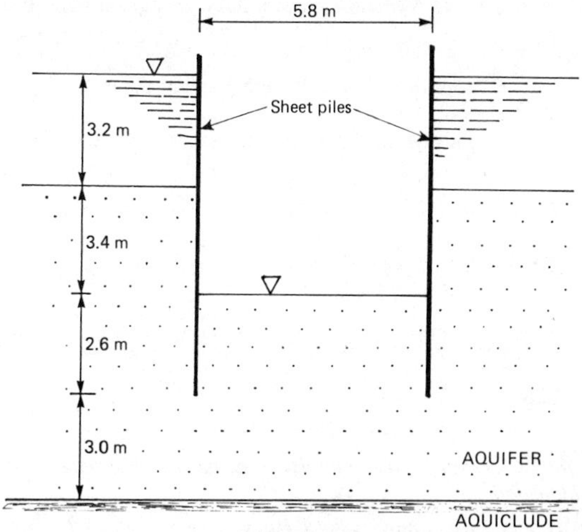

Fig 10.11

Hence $Q = kH \frac{N_f}{N}$

Hence $Q = kH \frac{N_f}{N_e}$

Worked example 10.6 Fig 10.10 shows the cross-section of a sheet-pile wall driven into a homogeneous stratum of sandy soil, which has a coefficient of permeability of 6.5×10^{-3} mm/s. By pumping, the water level on one side of the piles has been reduced from 4.5 m to 0.8 m. Draw a flow net for the seepage under the piling and from it estimate the seepage flow per hour.

A flow net for the seepage in this problem is shown in *Fig 10.10*.
The properties of this flow net are:

No. of flow channels, N_f	= 4.3
No. of equipotential drops, N_e	= 11
Total head drop, H	= 4.5 – 0.8

Then the seepage flow q $= kH \frac{N_f}{N_e}$

$$= 6.5 \times 10^{-6}(4.5 - 0.8) \times \frac{4.3}{11}$$

$$= 9.40 \times 10^{-6} \text{ m}^3/\text{s per metre}$$

$$= 9.40 \times \frac{60 \times 60}{10^6} = \mathbf{3.38 \times 10^{-2}\ m^3/h\ per\ metre}$$

Worked example 10.7 Determine the hydrostatic pressure at point P shown in *Fig 10.10*

There are 11 equal drops of pressure altogether and at point P, 3.1 of these 11 'units' are left.

The hydrostatic pressure at P during steady seepage will therefore be equal to

$\frac{3.1}{11} \times$ total head drop

$$u_p = \frac{3.1}{11}(4.5 - 0.8) = 1.04 \text{ m of water}$$

$$u_p = \mathbf{10.2\ kN/m^2}$$

TEST EXERCISES

COMPLETE QUESTIONS (answers on page 208)

Complete the following statements by inserting the most appropriate word or words in the spaces indicated.

1 The coefficient of permeability (k) is defined as the . . . . of flow induced by a hydraulic gradient of unity.

2 Darcy's law, provides the following equation for the quantity of water flowing in unit time:
$q =$ . . . .

3 A medium sand would normally have a coefficient of permeability equal to . . . . m/s.

4 The coefficient of permeability of a clay would normally be less than . . . . m/s.

5 The Constant Head permeability test is suitable for soils such as . . . . and . . . .

6 The Falling Head permeability test is suitable for soils such as . . . . and . . . .

7 In a Falling Head test the water levels in the standpipe are h_1 at time t_1 and h_2 at time t_2. The coefficient of permeability is then obtained from:

$$k = \frac{aL \log_e (\ldots\ldots\ldots\ldots)}{A (\ldots\ldots\ldots\ldots)}$$

8 The condition known as . . . . occurs when the hydraulic gradient becomes **critical**.

9 In terms of specific gravity (G_s) and void ratio (e), the critical hydraulic gradient is given by

$$i_c = \frac{\ldots\ldots\ldots\ldots}{\ldots\ldots\ldots\ldots}$$

10 The paths along which seepage takes place are called . . . .

11 In a flow net the set of lines representing equal fractions of the total head drop are called . . . .

12 In a flow net, impermeable boundaries are considered to be . . . . lines and permeable boundaries . . . . lines.

13 Flow lines and equipotential lines must cross each other at . . . .

14 If the number of flow channels is N_f and the number of equipotential drops is N_e, then the flow quantity represented by a flow net is given by
$A = k\,H$ . . . .

ESSAY QUESTIONS

1 Explain the terms **hydraulic gradient** and **coefficient of permeability** in the context of groundwater flow.

2 Compare the Constant Head and Falling Head permeability tests and comment on their suitability for particular soil types.

3 Discuss the advantages and disadvantages of laboratory permeability tests compared with field pumping tests.

4 Give an explanation of the **quick** or **quicksand** condition and suggest practical situations where it could occur.

CALCULATION EXERCISES (answers on page 208)

1 The following data was recorded during a constant head permeability test. From this calculate the coefficient of permeability k.

Diameter of sample	= 100 mm
Head difference between manometers	= 212 mm
Distance between manometer tappings	= 180 mm
Quantity of water collected in 90 s.	= 848 ml

2 A constant head permability test is to be conducted with a sample of diameter 75 mm on a sand having k values in the range 1×10^{-2} to 1×10^{-4} m/s. The steady state flow in each test is to be maintained at a head difference of 200 mm in a pair of manometers having tapping points 150 mm apart.

Calculate the maximum and minimum elapsed times required to collect 200 ml of water flowing through the permeameter.

3 In a falling head permeability test the following data was recorded. From this calculate the coefficient of permeability k .

Diameter of sample	= 75 mm
Length of sample	= 120 mm
Internal diameter of standpipe	= 6 mm
Initial level in standpipe	= 820 mm
Level in standpipe after 20 mins.	= 232 mm

4 In a field pumping test a pumping well was sunk through a layer of sand having a thickness of 8.8 m and underlain by a stratum of clay. The water table was initially 2.4 m below the horizontal ground level. At a steady-state pumping rate of 870 litres/min, the following observation-well data was recorded.

	Well A	*Well B*
Distance from pumping well (m):	18	40
Drawdown of water table (m):	2.38	1.17

Calculate the coefficient of permeability k.

5 A horizontal stratum of sand of 5.8 m thickness is overlain by a layer of clay with a horizontal surface and a thickness of 4.4 m. An impermeable layer underlies the sand layer. A pumping well was sunk through the clay and through to the bottom of the sand. Two observation wells were sunk through the clay just into the sand. At a steady-state pumping rate of 680 litres/min the following observations were recorded:

	Well A	*Well B*
Distance from pumping well (m)	15	35
Drawdown of piezometric surface (m)	2.56	1.78

Original piezometric surface level below ground surface = 1.8 m
Calculate to coefficient of permeability, k.

6 A confined sand aquifer is overlain by a horizontal layer of clay having a thickness of 4.5 m. If the piezometric surface level corresponding to the pore pressure at the base of the clay lies 1.8 m above the ground surface and the clay has a unit weight of 18.0 kN/m^3, calculate the depth to which an excavation could be taken in the clay before a bottom blow out (heave) would occur.

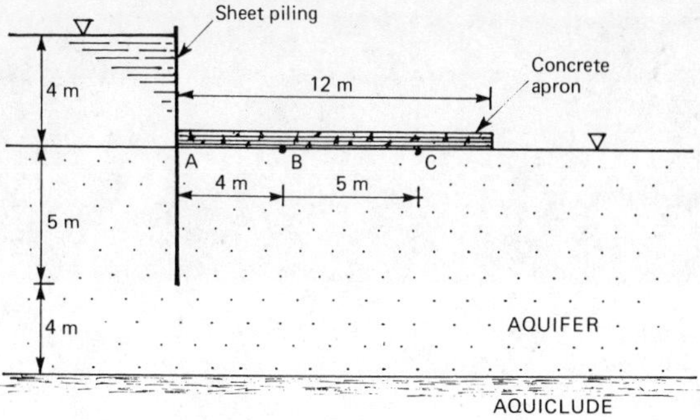

Fig 10.12

7 Sketch a flow net for the seepage under the line of sheet piling shown in *Fig 10.11* and from this calculate the estimated seepage loss per metre. ($k = 5 \times 10^{-6}$ m/s)

8 A sheet pile wall with a concrete apron is required to contain a head of water of 4.0 m as shown in *Fig 10.12*.

(a) If the coefficient of permeability of the sand is 84×10^{-6} m/s, draw a suitable flow net and from it give an estimate of the seepage loss.

(b) Calculate the hydrostatic uplift pressure acting on the concrete apron at points A, B and C.

11 Soil pressures and forces

11.1 TOTAL AND EFFECTIVE STRESS

Since water is an incompressible fluid virtually, it can transmit direct stress, but not shear stress. Shear stress has to be transmitted (or shear strength developed) through th granular fabric of a soil. If a mass of soil is saturated, the immediate effect of an increase in external stress is an equal increase in the **pore pressure**. This increase in pore pressure is dissipated if the porewater is permitted to drain away, the increase in applie stress is then transferred to the fabric of the soil as an increase in **effective** stress.

Pore pressure (u) the pressure induced in the porewater; also referred to as **neutral stress**, since it plays no part in the transmission of shear stress.

Effective stress (σ') the stress (pressure) transmitted through the soil fabric; the stress 'effective' in producing volume change or shear failure.

Total stress (σ) the **total** stress at a point is the sum of the pore pressure and the effecti stress at that point, i.e.

$$\sigma = \sigma' + u$$

Under static conditions (i.e. no seepage), the water table or piezometric surface represents the pore pressure datum. Consider the vertical stresses acting at a given depth z below the surface, as shown in *Fig 11.1*.

$$\begin{aligned}
\text{Total vertical stress, } \sigma_z &= \gamma_{sat} h + \gamma(z - h) \\
\text{Hydrostatic pore pressure, } u_z &= \gamma_w h \\
\text{Then, effective stress, } \sigma'_z &= \sigma_z - u_z \\
&= (\gamma_{sat} - \gamma_w)h + \gamma(z - h) \\
&= \gamma' h + \gamma(z - h)
\end{aligned}$$

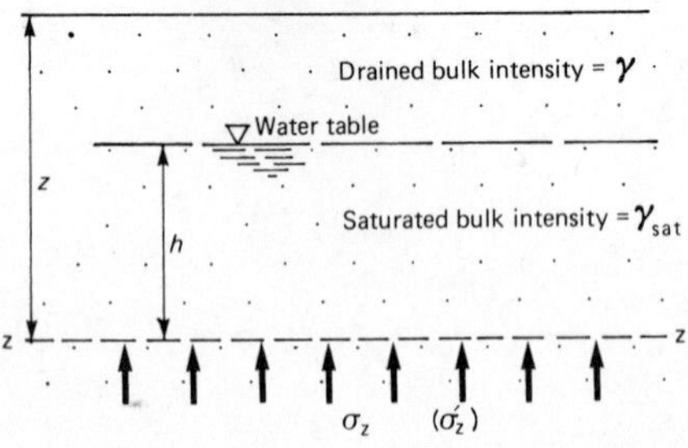

Fig 11.1

In fine-grained soils, the layer above the water table may be saturated due to capillary suction, in which case

$\sigma'_z = \gamma' h + \gamma_{sat}(z - h)$

If seepage is taking place and the vertical component of the seepage hydraulic gradient is i, the vertical seepage pressure will be $\pm i\gamma_w h$ (downwards = +ve)

Then $\sigma'_z = (\gamma' \pm i\gamma_w)h + \gamma_{sat}(z - h)$

Worked example 11.1 On a certain site the soil layers were found to be as follows:
0 – 3.5 m BS. Coarse sand (drained γ = 19.6 kN/m³, saturated γ = 20.4 kN/m³)
3.5 – 7.5 m BS. Silty clay (γ = 18.0 kN/m³)
Water table lies at a depth of 2.0 m
Draw an effective-stress/total-stress profile between 0 m and 7.5 m.

Above the WT: effective stress increment, $\Delta\sigma'_z = \gamma' z = \gamma z$
Below the WT: effective stress increment, $\Delta\sigma'_z = \gamma' z = (\gamma_{sat} - \gamma_w)z$

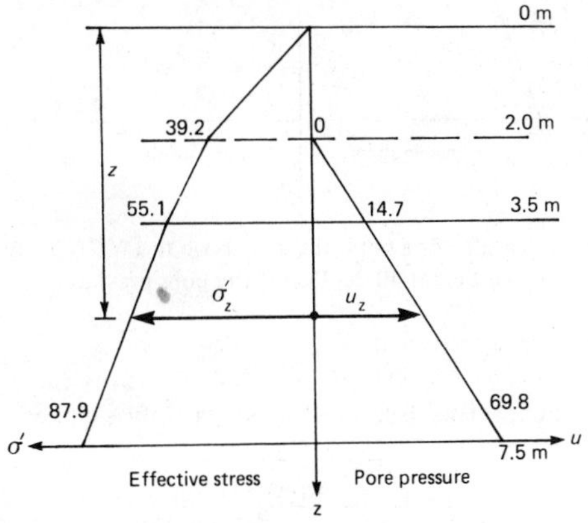

Fig 11.2

In a coarse-sand, the capillary suction is very low, so that the porewater pressure at all depths above the WT may be taken as zero; below the WT, $u_z = (z - 2.0)\,9.81$.

The calculations are tabulated below and the stress profiles plotted in *Fig 11.2*.

Depth BS (m)	*Stresses* (kN/m²) *Effective stress* $\Delta\sigma'_z$	σ'_z	*Pore pressure* u_z	*Total stress* σ_z
0		0	0	0
2.0	2.0 × 19.6 = 39.2	39.2	0	39.2
3.5	1.5 (20.4 – 9.81) = 15.9	55.1	1.5 × 9.81 = 14.7	69.8
7.5	4.0 (18.0 – 9.81) = 32.8	87.9	5.5 × 9.81 = 53.9	141.8

Worked example 11.2 The soil layers on a certain site were found to be as follows:
0 – 4 m Silty fine sand ($\gamma = 19.2\ kN/m^3$)
4 – 8 m Firm brown clay ($\gamma = 18.1\ kN/m^3$)
Water table lies at a depth of 2.2 m
Draw an effective-stress/total-stress profile from 0 m to 8 m.

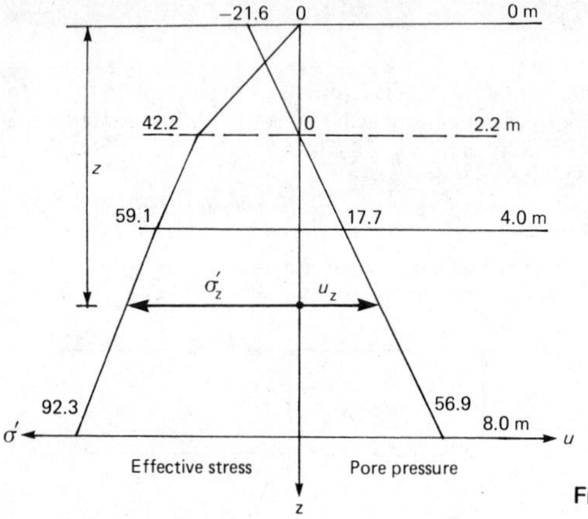

Fig 11.3

Since the upper layer is a silty fine sand, the soil above the WT is likely to be saturated with capillary moisture, so that at all depths (z) the pore pressure,

$u_z = (2.2 - z)\,9.81.$

Above WT: effective stress increment, $\Delta\sigma'_z = \gamma' z = \gamma z$

effective stress increment, $\Delta\sigma'_z = \gamma' z = (\gamma_{sat} - \gamma_w)z$

The calculations are tabulated below and the stress profiles plotted in *Fig 11.3.*

Depth BS (m)	*Stresses* (kN/m^2)			
	Effective stress		*Pore pressure* u_z	*Total stress* σ_z
	$\Delta\sigma'_z$	σ'_z		
0		0	−2.2 × 9.81 = 21.6	−21.6
2.2	2.2 × 19.2 = 42.2	42.2	0	42.2
4.0	1.8(19.2 – 9.81) = 16.9	59.1	1.8 × 9.81 = 17.7	76.8
8.0	4.0(18.1 – 9.81) = 33.2	92.3	5.8 × 9.81 = 56.9	149.2

11.2 VERTICAL STRESS DUE TO SURCHARGE LOADING

A load applied to the surface or just below the surface will induce an associated vertical stress at other points in a soil mass. Generally, for a **surcharge pressure** (q),

$\sigma'_z = Iq$

TABLE 11.1 Influence factor (I_R) for vertical stress under a corner of a uniformly-loaded rectangle

B/Z \ L/Z	0.1	0.2	0.3	0.4	0.5	0.6	0.7	0.8	0.9	1.0	1.4	2.0	3.0	5.0	∞
0.1	0.0047	0.0092	0.0132	0.0168	0.0198	0.0222	0.0242	0.0258	0.0270	0.0279	0.0301	0.0311	0.0315	0.0316	0.0316
0.2	0.0092	0.0179	0.0259	0.0328	0.0387	0.0435	0.0474	0.0504	0.0528	0.0547	0.0589	0.0610	0.0618	0.0620	0.0620
0.3	0.0132	0.0259	0.0374	0.0474	0.0560	0.0630	0.0686	0.0731	0.0766	0.0794	0.0856	0.0887	0.0898	0.0901	0.0902
0.4	0.0168	0.0328	0.0474	0.0602	0.0711	0.0801	0.0873	0.0931	0.0977	0.1013	0.1094	0.1134	0.1150	0.1154	0.1154
0.5	0.0198	0.0387	0.0560	0.0711	0.0840	0.0947	0.1034	0.1104	0.1158	0.1202	0.1300	0.1350	0.1368	0.1374	0.1375
0.6	0.0222	0.0435	0.0629	0.0801	0.0947	0.1069	0.1168	0.1247	0.1310	0.1361	0.1475	0.1533	0.1555	0.1561	0.1562
0.7	0.0240	0.0474	0.0686	0.0873	0.1034	0.1168	0.1277	0.1365	0.1436	0.1491	0.1620	0.1686	0.1711	0.1719	0.1720
0.8	0.0258	0.0504	0.0731	0.0931	0.1104	0.1247	0.1365	0.1461	0.1537	0.1598	0.1739	0.1812	0.1841	0.1849	0.1850
0.9	0.0270	0.0528	0.0766	0.0977	0.1158	0.1311	0.1436	0.1537	0.1619	0.1684	0.1836	0.1915	0.1947	0.1956	0.1958
1.0	0.0279	0.0547	0.0794	0.1013	0.1202	0.1361	0.1491	0.1598	0.1684	0.1752	0.1914	0.1999	0.2034	0.2044	0.2046
1.4	0.0301	0.0589	0.0856	0.1094	0.1300	0.1475	0.1620	0.1739	0.1836	0.1914	0.2102	0.2206	0.2250	0.2263	0.2266
2.0	0.0311	0.0610	0.0887	0.1134	0.1350	0.1533	0.1686	0.1812	0.1915	0.1999	0.2206	0.2325	0.2378	0.2395	0.2399
3.0	0.0315	0.0618	0.0898	0.1150	0.1368	0.1555	0.1711	0.1841	0.1947	0.2034	0.2250	0.2378	0.2420	0.2461	0.2465
5.0	0.0316	0.0620	0.0901	0.1154	0.1374	0.1561	0.1719	0.1849	0.1956	0.2044	0.2263	0.2395	0.2461	0.2486	0.2491
∞	0.0316	0.0620	0.0902	0.1154	0.1375	0.1562	0.1720	0.1850	0.1958	0.2046	0.2266	0.2399	0.2465	0.2492	0.2500

$\sigma_z = qI_R$

The value of I is obtained from a geometrical analysis encompassing the shape of the loaded area, the distribution of loading and the possition of the point in question. *Table 11.1* gives the influence factors required to calculate σ'_z under a corner of a uniformly-loaded rectangle. Similar tables (not given here) may be deduced for concentrated loads, circular loads and strip loads of different types.

When the reference point is not at a corner of the loaded area, four rectangles are drawn each having a corner at the point and the principle of superposition applied.

Worked example 11.3 The plan of a rectangular foundation is shown in *Fig 11.4*. Using the influence factors given in *Table 11.1*, calculate the vertical direct stress induced by a uniform surcharge pressure of 150 kN/m² at the following points: (a) 8 m below point A; (b) 4 m below point B.

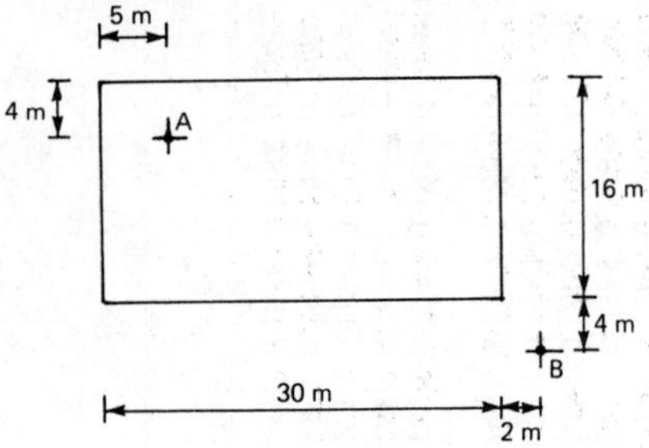

Fig 11.4

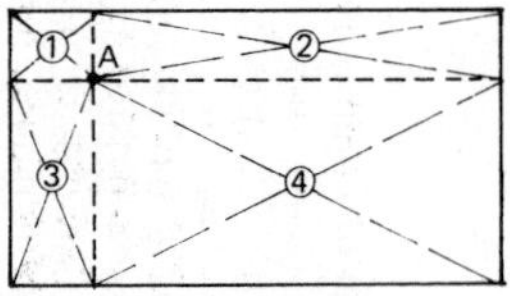

Fig 11.5

(a) First, divide the foundation into four component rectangles (1, 2, 3, 4), each having a corner at A (*Fig 11.5*). The vertical stress below A is then the sum of the stresses induced by the component rectangular loaded areas.

$$\sigma_{z(A)} = \sigma_{z(1)} + \sigma_{z(2)} + \sigma_{z(3)} + \sigma_{z(4)}$$
$$= q\,(I_1 + I_2 + I_3 + I_4)$$

The calculations are tabulated below:

$z = 8$ m

Rectangle	See Table 11.1		
	L/z	B/z	I
1	5/8 = 0.625	4/8 = 0.5	0.0969
2	25/8 = 2.125	12/8 = 1.5	0.2277
3	5/8 = 0.625	4/8 = 0.5	0.1522
4	25/8 = 2.125	4/8 = 0.5	0.1368

Then $\sigma_{z(A)} = 150\,(0.0969 + 0.2277 + 0.1522 + 0.1368)$
$= \mathbf{92\ kN/m^2}$

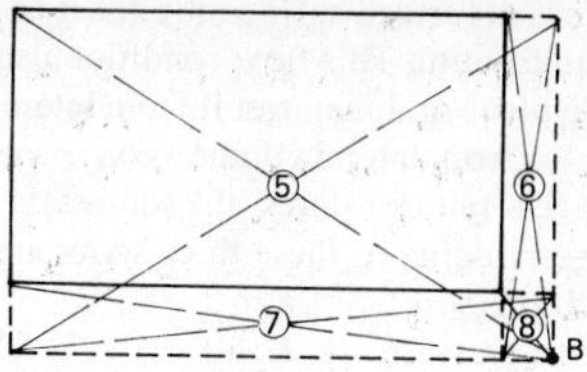

Fig 11.6

(b) Again four rectangles (5, 6, 7, 8) are arranged so that each has a corner at point B (*Fig 11.6*). The vertical stress below B is then

$$\sigma_{z(B)} = \sigma_{z(5)} - \sigma_{z(6)} - \sigma_{z(7)} + \sigma_{z(8)}$$
$$= q\,(I_5 - I_6 - I_7 + I_8)$$

The calculations are tabulated below.

$z = 4$ m

Rectangle	*See Table 11.1*		
	L/Z	*B/Z*	*I*
5	32/4 = 8.0	20/4 = 5.0	0.2490
6	2/4 = 0.5	20/4 = 5.0	0.1374
7	32/4 = 8.0	4/4 = 1.0	0.2044
8	2/4 = 0.5	4/4 = 1.0	0.1202

Then $\sigma_{z(B)}$ = 150 (0.2490 – 0.1374 – 0.2044 + 0.1202) = **4 kN/m²**

11.3 LATERAL EARTH PRESSURE

The pressure at a given depth in a liquid acts equally in all directions. In soils, which possess internal shear strength, the lateral stress is not often equal to the vertical stress at the same point, although it still remains a function of it. The relationship between lateral and vertical stress depends on whether the soil is behaving elastically or

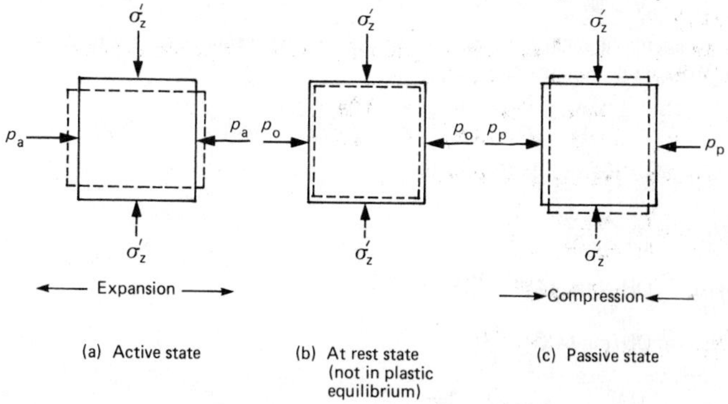

Fig 11.7 State of plastic equilibrium

plastically. When a soil mass is deformed sufficiently for it to flow, it is said to have reached a state of **plastic equilibrium**. This flow condition usually takes the form of a shear failure along a definite plane and may result from **lateral expansion**, when it is said to be in an **active state**, or from **lateral compression**, producing a **passive state**. Between these two limiting (i.e. failure) states, the soil is said to be **at rest** (*Fig 11.7*).

The lateral pressures corresponding to these three states are as follows

p_a = **active** lateral earth pressure = $K_a \sigma'_z$
p_p = **passive** lateral earth pressure = $K_p \sigma'_z$
p_o = lateral earth pressure **at rest** = $K_o \sigma'_z$

K_a, K_p and K_o are known as **earth pressure coefficients**. Remember that p_a and p_p are lateral stresses at a point of shear failure, but p_o occurs when the lateral strain is virtually zero.

EARTH PRESSURE AT REST ($p_o = K_o \sigma'_z$)

The value of K_o may be deduced theoretically from elastic theory: $K_o = \dfrac{\nu}{1-\nu}$.

Where ν = Poisson's ratio for the soil. For a saturated clay, $\nu = 0.5$ and $K_o = 1.0$, so that the soil behaves elastically as a liquid. For sand, ν ranges between 0.2 and 0.5, so that K_o ranges between 0.25 and 0.5. K_o may also be estimated empirically if the angle of shearing resistance (ϕ') has been determined.

$$K_o \doteq 1 - \sin \phi'$$

RANKINE'S THEORY FOR ACTIVE AND PASSIVE PRESSURE

In Rankine's theory (presented in 1857) a homogeneous cohesionless soil mass having a horizontal surface is considered. *Fig 11.8* shows the combined Mohr circle construction for the active and passive states:

ACTIVE STATE: major principal stress = σ'_z
minor principal stress = p_a = $K_a \sigma'_z$

PASSIVE STATE: major principal stress = p_p = $K_a \sigma'_z$
minor principal stress = σ'_z

(Note: $p_a < \sigma'_z < p_p$)

The coefficients of earth pressure (K_a and K_p) can be evaluated from the geometry of the Mohr circle construction as follows:

$$K_a = \frac{p_a}{\sigma'_z} = \frac{OA}{OB} = \frac{OF - AF}{OF + FB} = \frac{1 - AF/OF}{1 + FB/OF}$$

But, AF = FB = FD and FD/OF = $\sin \phi'$

$$\text{Then } K_a = \frac{1 - \sin \phi'}{1 + \sin \phi'}$$

$$\text{Also OA} = \text{OH} \times \tan\left(45° - \frac{\phi'}{2}\right)$$

$$\text{and OB} = \text{OH}/\tan\left(45° - \frac{\phi'}{2}\right)$$

$$\text{Then } K_a = \frac{OA}{OB} = \tan^2\left(45° - \frac{\phi'}{2}\right)$$

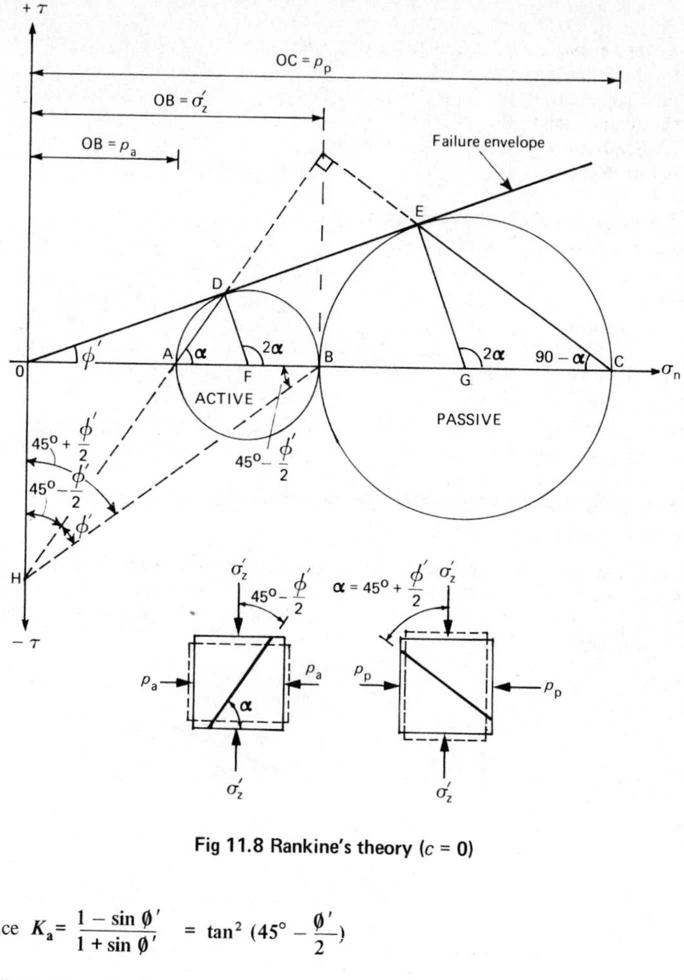

Fig 11.8 Rankine's theory (c = 0)

Hence $K_a = \dfrac{1 - \sin \phi'}{1 + \sin \phi'} = \tan^2 (45° - \dfrac{\phi'}{2})$

Similarly, it can be shown that

$$K_p = \frac{1 + \sin \phi'}{1 - \sin \phi'} = \tan^2 (45° + \frac{\phi'}{2}) = \frac{1}{K_a}$$

The angle between the plane of shear failure and the direction of σ_z', i.e. vertical, is $(45° - \dfrac{\phi'}{2})$ in the active state, and $(45° + \dfrac{\phi'}{2})$ in the passive state.

Worked example 11.4 Evaluate the coefficient of earth pressure K_o, K_a and K_p for the following soils.
(a) Coarse gravel: $\phi' = 42°$
(b) Medium sand: $\phi' = 30°$
(c) Sandy silt: $\phi' = 18°$
(d) Firm clay: $\phi_u = 8°$

The calculations are tabulated below.

Soil	ϕ' or ϕ_u	*At rest* $K_o = 1 - \sin\phi'$	*Active* $K_a = \frac{1-\sin\phi'}{1+\sin\phi'}$	*Passive* $K_p = \frac{1+\sin\phi'}{1-\sin\phi'}$
a	42°	0.331	0.198	5.04
b	30°	0.500	0.333	3.00
c	18°	0.691	0.528	1.89
d	8°	0.861	0.756	1.32

SOLUTION FOR COHESIVE SOILS

Rankine's theory (1857) dealt only with cohesionless granular materials such as sand, gravel, grain, etc. In 1915, Bell produced an adaptation of Rankine's analysis incorporating the effect of cohesion. *Fig 11.9* shows the Mohr circle and failure envelope corresponding to a c – ϕ soil.

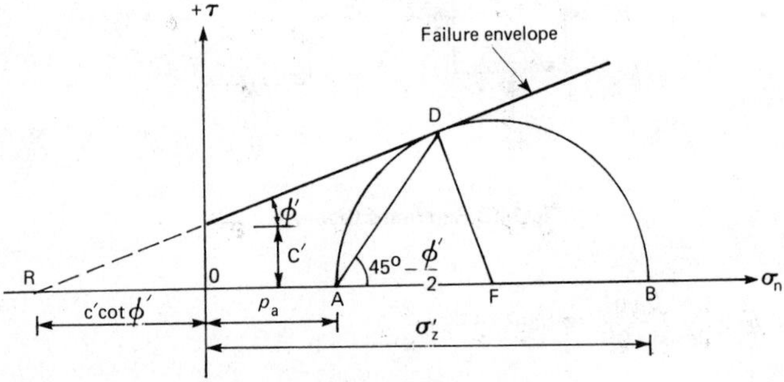

Fig 11.9 Solution for cohesive soil

From the diagram:
$$\sin\phi' = \frac{\tfrac{1}{2}(\sigma'_z - p_a)}{c'\cot\phi' + p_a + \tfrac{1}{2}(\sigma'_z - p_a)}$$

$$= \frac{\sigma'_z - p_a}{2c'\cot\phi' + \sigma'_z + p_a}$$

Rearranging:
$$p_a + \sin\phi' p_a = \sigma'_z - \sin\phi'\sigma'_z - 2c'\cot\phi'\sin\phi'$$

So that
$$p_a = \sigma'_z\,\frac{1-\sin\phi'}{1+\sin\phi'} - \frac{2c'\cos\phi'}{1+\sin\phi'}$$

But $\cos \phi' = \sqrt{1 - \sin^2 \phi'} = \sqrt{(1 - \sin \phi')(1 + \sin \phi')}$

Therefore

$$p_a = \sigma'_z \frac{1 - \sin \phi'}{1 + \sin \phi'} - 2c' \sqrt{\frac{1 - \sin \phi'}{1 + \sin \phi'}}$$

$$= Ka\, \sigma'_z - 2c' \sqrt{K_a}$$

Hence the active lateral earth pressure in a cohesive soil is given by

$$p_a = K_a \sigma'_z - 2c' \sqrt{K_a}$$

Similarly the passive pressure is given by

$$p_p = K_p \sigma'_z + 2c' \sqrt{K_p}$$

These two general equations may be referred to as the *Rankine/Bell equations.*

Worked example 11.5 Calculate the active lateral earth pressure at a depth of 5 m beneath a horizontal surface in the following drained homogeneous soils.

(a) $\phi' = 32°$	$c' = 0$	$\gamma = 20$	kN/m^3
(b) $\phi' = 25$	$c' = 0$	$\gamma = 18.5$	kN/m^3
(c) $\phi' = 18°$	$c' = 20\ kN/m^2$	$\gamma = 18.0$	kN/m^3
(d) $\phi_u = 8°$	$c_u = 50\ kN/m^2$	$\gamma = 18.0$	kN/m^3
(e) $\phi_u = 0$	$c_u = 70\ kN/m^2$	$\gamma = 18.0$	kN/m^3

(a) $K_a = \dfrac{1 - \sin 32°}{1 + \sin 32°} = 0.307 \qquad \sigma'_z = 20 \times 5 = 100\ kN/m^2$

$$= Ka\, \sigma'_z - 2c' \sqrt{Ka}$$
$$= 0.307 \times 100 = \mathbf{30.7\ kN/m^2}$$

(b) $K_a = \dfrac{1 - \sin 25°}{1 + \sin 25°} = 0.406 \qquad \sigma'_z = 18.5 \times 5 = 92.5\ kN/m^2$

$$p_a = K_a \sigma'_z - 2c' \sqrt{K_a}$$
$$= 0.406 \times 92.5 = \mathbf{37.6\ kN/m^2}$$

(c) $K_a = \dfrac{1 - \sin 18°}{1 + \sin 18°} = 0.691 \qquad \sigma'_z = 18.0 \times 5 = 90.0\ kN/m^2$

$$p_a = K_a \sigma'_z - 2c' \sqrt{K_a}$$
$$= 0.691 \times 90.0 - 2 \times 20 \sqrt{0.691} = \mathbf{28.9\ kN/m^2}$$

(d) For the undrained case, $K_a \doteq \dfrac{1 - \sin \phi_u}{1 + \sin \phi_u} = \dfrac{1 - \sin 8°}{1 + \sin 8°} = 0.756$

and $p_a = K_a \sigma_z - 2c_u \sqrt{K_a}$
$$= 0.756 \times 18 \times 5 - 2 \times 50 \sqrt{0.756} = \mathbf{18.9\ kN/m^2}$$

(e) When $\phi_u = 0$, $K_a = 1.0$ and $\sqrt{K_a} = 1.0$

$$p_a = \sigma_z - 2c_u$$
$$= 18 \times 5 - 2 \times 70 = \mathbf{-50.0\ kN/m^2}$$

One of the principal applications of earth pressure theory is in the solution of problems relating to the stability and design of earth retaining walls. For this purpose, it is necessary to translate the loading conditions and soil parameters into a **lateral thrust** (*P*) which is the resultant force acting on the wall due to earth pressure.

The distribution of earth pressure against the supporting face is first of all determined and the resultant lateral thrust calculated as the area of the diagram. In the worked examples that follow the following conditions are considered:

Active pressure:	cohesionless soil	– drained and waterlogged, with and without surface surcharge.
	cohesive soil	– no surcharge
Passive pressure:	cohesionless soil	– drained
At rest:	cohesionless soil	– drained

Worked example 11.6 A retaining wall having a smooth vertical back is to retain a cohesionless soil with a horizontal surface to a depth of 12 m. The soil has the following properties: $\phi' = 32°$, drained $\gamma = 19.2$ kN/m², saturated $\gamma = 20.0$ kN/m³. Determine the active lateral thrust induced on the wall when:
(a) the soil is fully drained;
(b) the water table stands at a depth of 6 m;
(c) the ground is fully waterlogged.

$$K_a = \frac{1 - \sin 32°}{1 + \sin 32°} = 0.307$$

(a) When the soil is fully drained, then at any depth, γ' = drained γ, and therefore $\sigma'_z = 19.2\,z$

At the base of the wall,

$\sigma'_z = 19.2 \times 12$

and $p_a = 0.307 \times 19.2 \times 12 = 70.7$ kN/m²

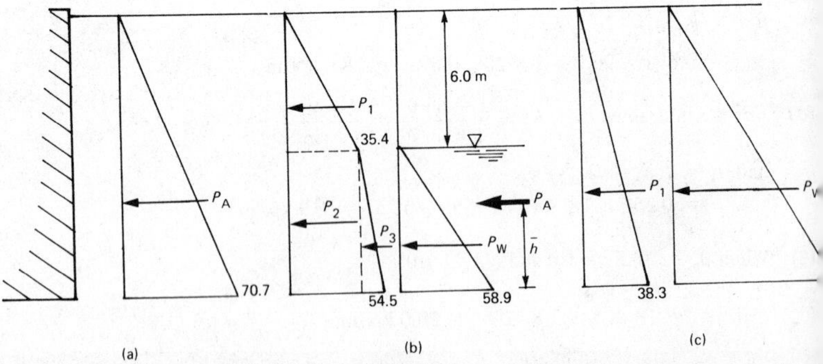

Fig 11.10

The pressure distribution is triangular as shown in *Fig 11.10(a)*, with the force P_A being equal to the area of the pressure distribution diagram.

$P_A = ½ \times 12 \times 70.7 =$ **424 kN/m**

acting at a height of **4.0 m above the base.**

(b) Above the WT, $\gamma' =$ drained $\gamma = 19.2 \text{ kN/m}^3$

Below the WT, $\gamma' = \gamma_{sat} - \gamma_w = (20.2 - 9.81) \text{ kN/m}^3$

At $z = 6$ m $\quad p_a = 0.307 \times 19.2 \times 6 = 35.4 \text{ kN/m}^2$

At $z = 12$ m $\quad p_a = 35.4 + 0.307(20.2 - 9.81)6 = 54.5 \text{ kN/m}^2$

Then the pore pressures will be:

Between $z = 0$ and $z = 6$ m, $\quad u_w = 0$

At $z = 12$ m, $\quad u_w = 9.81 \times 6 = 58.9 \text{ kN/m}^2$

The pressure distribution is shown in *Fig 11.10(b)*, giving the followi ng resultants:

$$P_1 = ½ \times 6 \times 35.4 = 106.2 \text{ kN/m}$$
$$P_2 = 6 \times 35.4 = 212.4 \text{ kN/m}$$
$$P_3 = ½ \times 6\,(54.5 - 35.4) = 57.3 \text{ kN/m}$$
$$P_w = ½ \times 6 \times 58.9 = 176.7 \text{ kN/m}$$

The total thrust on the wall, $P_A = P_1 + P_2 + P_3 + P_w =$ **553 kN/m**

The position of P_A may be obtained by taking moments:

$$P_A \times \bar{h} = P_1(6 + 6/3) + P_2 \times 3 + P_3 \times 6/3 + P_w \times 6/3$$
$$= 849.6 + 637.2 + 114.6 + 530.1 = 2132$$

Then $\bar{h} = \dfrac{2132}{553} =$ **3.85 m**

(c) When fully waterlogged (i.e. WT at surface) $\gamma' = \gamma_{sat} - \gamma_w$ at all depths.

At $z = 12$ m, $p_a = 0.307\,(20.2 - 9.81) \times 12 = 38.3 \text{ kN/m}^2$

$p_w = 9.81 \times 12 = 117.7 \text{ kN/m}^2$

The pressure distribution is shown in *Fig 11.10(c)* giving the following resultants:

$$P_1 = ½ \times 12 \times 38.3 = 229.8 \text{ kN/m}$$
$$P_w = ½ \times 12 \times 117.7 = 706.2 \text{ kN/m}$$

The total thrust on the wall, $P_A = P_1 + P_w =$ **936 kN/m**

Worked example 11.7 A retaining wall having a smooth vertical back is to retain a drained cohesionless soil with a horizontal surface to a depth of 9 m. The soil has the following properties: $\phi' = 30°$, $\gamma = 19.8 \text{ kN/m}^3$. Determine the total active thrust acting on the wall when:

(a) there is no surface surcharge;

(b) there is a uniform surcharge on the soil surface of 50 kN/m^2.

$$K_a = \frac{1 - \sin 30°}{1 + \sin 30°} = 0.333$$

(a) At $z = 9$ m $\quad p_a = K_a\,\sigma'_z$

$= 0.333 \times 19.8 \times 9 = 59.3 \text{ kN/m}^2$

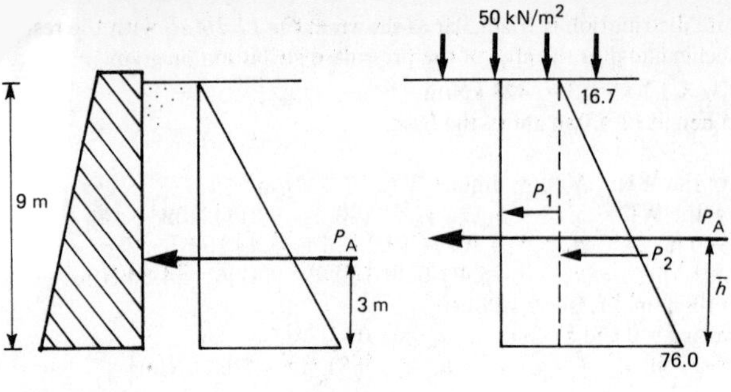

Fig 11.11

The pressure distribution is shown in *Fig 11.11(a).*
Total thrust, $P_A = \frac{1}{2} \times 9 \times 59.3 =$ **267 kN/m**
Acting **3 m above the base.**

(b) The surface surcharge may be considered to be transmitted uniformly to all depths,
Then $\sigma'_z = 50 + \gamma' z$

at $z = 0$, $p_a = K_a \sigma'_z = 0.333(50 + 19.8 \times 0) = 16.7$ kN/m²
at $z = 9$ m, $p_a = K_a \sigma'_z = 0.333(50 + 19.8 \times 9) = 76.0$ kN/m²

The pressure distribution is shown in *Fig 11.11(b)* giving the following resultants:

$P_1 = 16.7 \times 9 = 150.3$ kN/m
$P_2 = \frac{1}{2}(76.0 - 16.7)\,9 = 266.9$ kN/m
Total thrust, $P_A = P_1 + P_2 =$ **417 kN/m**

Acting at $\bar{h}$ above the base

$$\bar{h} = \frac{150.3 \times 4.5 + 266.9 \times 3}{417} = \mathbf{3.54\ m}$$

Worked example 11.8 A retaining wall with a smooth vertical back is to retain a cohesive soil with a horizontal unsurcharged surface. The soil has the following properties:
$c' = 20$ kN/m² $\quad \phi' = 15°$ $\quad \gamma = 18.0$ kN/m³

Calculate the lateral active thrust on the wall due to a soil depth of 12 m.

$$K_a = \frac{1 - \sin 15°}{1 + \sin 15°} = 0.589$$

At $z = 0$, $\sigma'_z = \gamma' z = 0$
$p_a = K_a \sigma'_z - 2c' \sqrt{K_a}$
$= 0 - 2 \times 20 \times \sqrt{0.589} = -30.7$ kN/m³
At $z = 12$ m, $\sigma'_z = 18.0 \times 12 = 216$ kN/m²
$p_a = 0.589 \times 216 - 30.7 = 96.5$ kN/m³

The pressure distribution is shown in *Fig 11.12.*

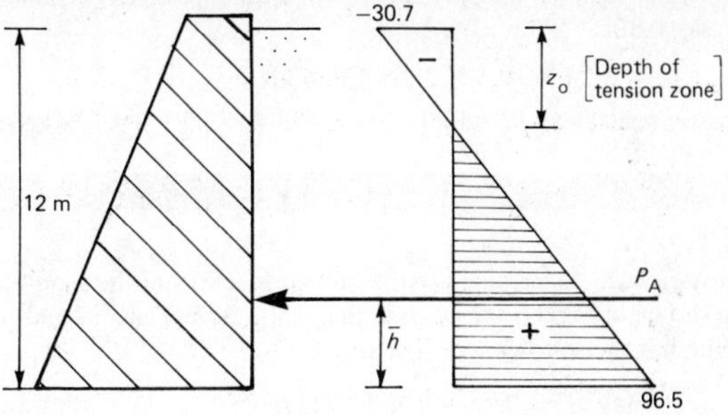

Fig 11.12

The negative pressure near the surface is due to the cohesion, but this is only an internal effect – it is *not* transmitted to the wall. Therefore, only the area of the positive part of the pressure distribution is taken as providing the active thrust on the wall.

The depth of the tension zone (z_o) may be found by similar triangles.

$$\frac{z_o}{30.7} = \frac{12}{30.7 + 96.5} \quad \therefore z_o = \frac{12 \times 30.7}{127.2} = 2.90 \text{ m}$$

The active thrust, P_A = shaded area
= ½ (12.0 – 2.90) 96.5 = **439 kN/m**

Acting at $\bar{h}$ above the base

$\bar{h} = \frac{1}{3}(12.0 - 2.9) =$ **3.03 m**

Worked example 11.9 Calculate the resultant passive resistance (P_p) offered by the soil in front of a retaining wall as shown is *Fig 11.13*. The soil properties are:

$\phi' = 32°$ $\quad \gamma = 19.5$ kN/m $\quad c' = 0$

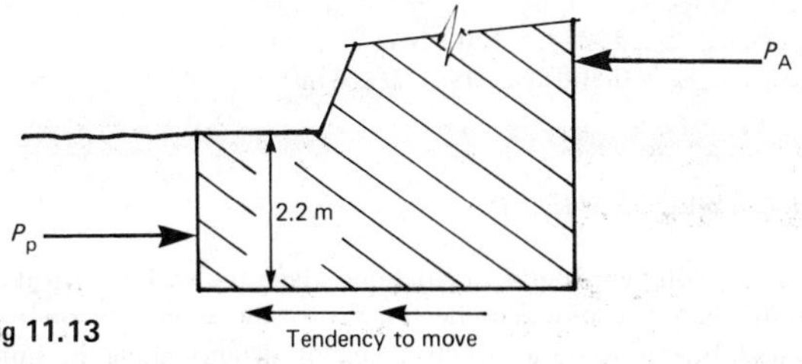

Fig 11.13

$$K_p = \frac{1 + \sin 32°}{1 - \sin 32°} = 3.25$$

The appropriate Rankine/bell expression is

$$p_p = K_p\, \sigma_z' + 2c'\sqrt{K_p}$$

However, it is usual to apply a safety factor when using passive resistance in stability calculations, say 0.80.

At $z = 2.2$ m, $p_p = 3.25 \times 19.5 \times 2.2 = 139.4$ kN/m^2

Resultant passive resistance, $P_p = 0.80 \times \frac{1}{2} \times 139.4 \times 2.2 + 0 =$ **123 kN/m**

Worked example 11.10 Calculate the pressures acting on the rigid buried structure shown in *Fig 11.14*. Assume that soil is homogeneous and fully drained and has the following properties:

$c' = 0$ $\quad \phi' = 30°$ $\quad \gamma = 20$ kN/m^3

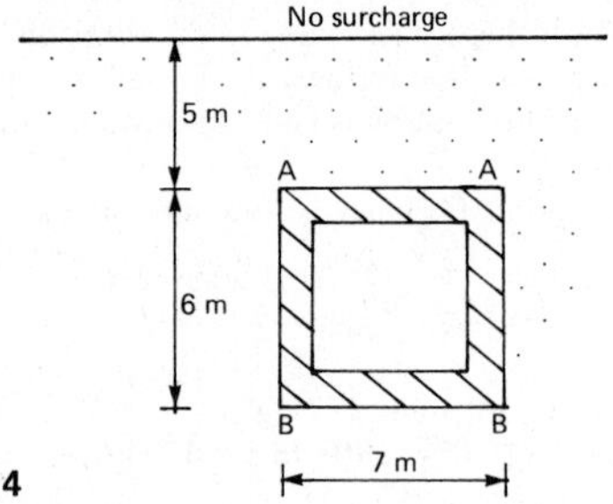

Fig 11.14

Since the structure is rigid, no yielding will take place, thus **at-rest** conditions prevail.

$K_o = 1 - \sin 30° = 0.500$

At level AA ($z = 5$m)

Vertical stress, $\sigma'_z = 20 \times 5 =$ **100 kN/m²**

Horizontal stress, $p_o = K_o \sigma'_z$

$= 0.500 \times 100 =$ **50 kN/m²**

At level BB ($z = 11$m)

Vertical stress, $\sigma'_z = 20 \times 11 =$ **220 kN/m²**

Horizontal stress, $p_o = 0.500 \times 220 =$ **110 kN/m²**

11.5 COULOMB'S WEDGE THEORY

Rankine's theory is easily applied to calculations where the wall is vertical and the surface horizontal, but it becomes cumbersome when the boundary conditions are more complicated. In addition, the effects of shear resistance along the supporting surface are ignored. The **wedge theory** proposed by Coulomb (1776), may be applied in more complex cases and produces lower solutions by allowing for friction and/or adhesion between the soil and the wall.

In considering an **active** failure, it is assumed that the soil behind the wall will shear along a plane AC (*Fig 11.15*) displacing a wedge of soil ABC. At the moment

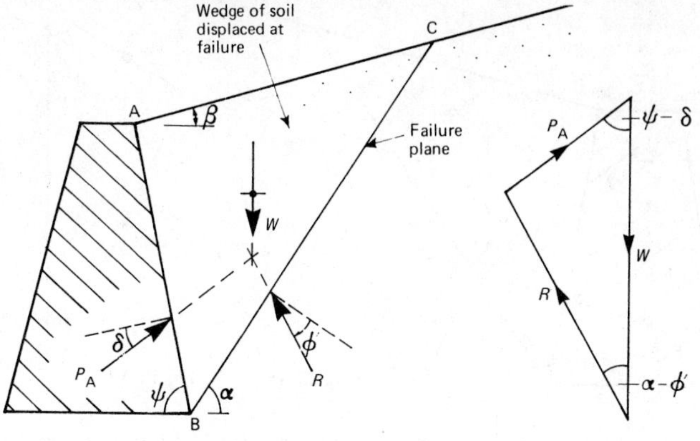

(a) Cross-section (b) Triangle of forces

Fig 11.15 Coulomb's wedge theory

immediately prior to failure the wedge will be held in static equilibrium by three forces (when $c = 0$), which are:

the weight of the wedge ABC	$= W$
the reaction from the soil below the failure plane	$= R$
the reaction from the wall (resultant thrust)	$= P_A$

The frictional resistance along the failure plane is represented by the displacement of R at an angle equal to the angle of shearing resistance (ϕ'). Similarly, P_A is displaced by the *angle of wall friction* (δ). The triangle of forces can be drawn when the wedge weight W and the angles ($\psi - \delta$) and ($\alpha - \phi'$) have been calculated, and thence a value for P_A obtained. Since the angle of the failure plane (α) is initially unknown, a series of trial wedges is selected and the maximum (and thus critical) value of P_A obtained together with the corresponding critical angle (α_f).

Worked example 11.11 Use Coulomb's wedge theory to evaluate the magnitude of the active thrust (P_A) due to the trial wedge shown in *Fig 11.16*. The soil properties are $c' = 0$, $\phi' = 30°$, $\gamma = 20$ kN/m³, $\delta = 20°$.

By drawing the cross-section to scale, the dimensions AB and AC are determined:

AB = 12.2 m
AC = 13.4 m

Then the area of wedge ABC $= \frac{1}{2} \sin 95° \times AB \times AC$
$= \frac{1}{2} \sin 95° \times 12.2 \times 13.4$
$= 81.43\ m^2$

The weight of the wedge, $W = 20 \times 81.43$
$= 163$ kN/m

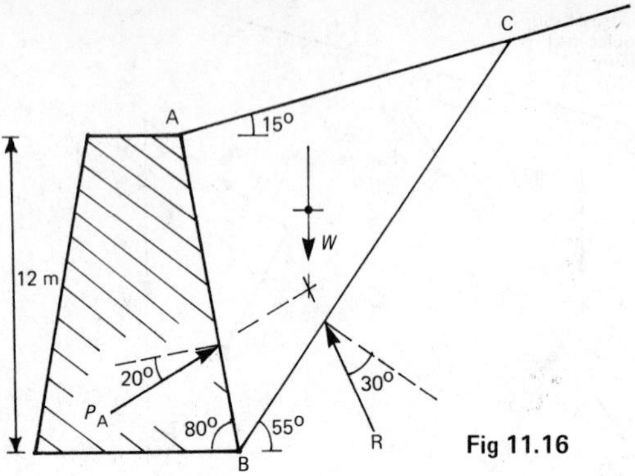

Fig 11.16

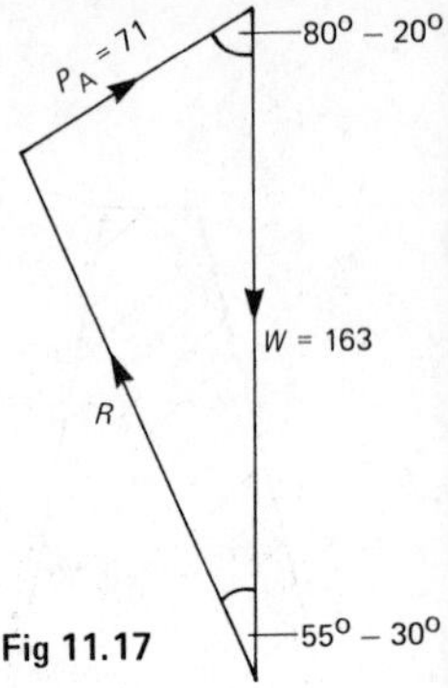

Fig 11.17

The triangle of forces may now be drawn as shown in *Fig 11.17*, from which the magnitude of P_A is scaled off:
For this trial wedge, P_A = **71 kN/m**

11.6 FACTORS AFFECTING THE STABILITY OF SLOPES

The principal forces tending to cause instability in either natural or man-made slopes are those due to gravity and seepage. Resistance to failure depends partly on the shear strength of the rock or soil and partly on the geometry of the slope. A naturally degraded slope will often be in a **just-stable** (factor of safety = 1.0) state, so that any small adverse change in conditions may precipitate failure. Man-made slopes will usually have a calculated factor of safety of 1.25 – 1.5 at the time of construction, but here again changes in conditions may reduce this to the point of failure.

Fig 11.18 Slope stability

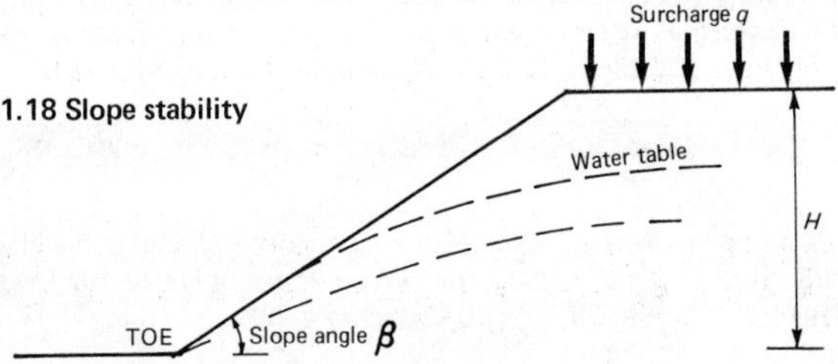

In general, a slope will become *less stable* as a result of the following changes (see *Fig 11.18*):

increase in slope angle (β)
increase in slope height (H)
increase in surcharge at top of slope
increase in soil density } due perhaps to an *increase* in moisture content
decrease in cohesion }

decrease in angle of shearing resistance
rise in water table level, i.e. waterlogging
removal of material from the toe
removal of stabilising vegetation

Four different types of failure movement can be identified (*Fig 11.19*):

FALLS Movement away from discontinuities such as joints, fissures, bedding planes, etc.; typically occurring in rocks.
TRANSLATIONAL SLIDES Movement of a shallow weak layer parallel to the slope surface.
ROTATIONAL SLIPS Movement along a curved shear surface, typically occurring in cohesive soils.
FLOWS Occur in weak saturated soils due to increase in pore pressure and loss of shear strength, the soil becoming liquid.

TEST EXERCISES

COMPLETION QUESTIONS (answers on page 208)

Complete the following statements by inserting the most appropriate word or words in the spaces indicated.

1 Pore pressure is transmitted through the . . . .

2 Effective stress is transmitted through the . . . .

3 Write down (using symbols) the relationship between total stress (σ), effective stress (σ') and pore pressure (u): . . . .

4 With the water table standing at the surface of a homogeneous soil, the effective vertical overburden stress at a depth z is given by: σ'_z = ...

5 Write down the symbols which indicate the coefficients of active, passive and at-rest earth pressure:
(i) . . . . (ii) . . . . (iii) . . . .

6 Write down the relationships used to give the coefficient of earth pressure at rest (i) theoretically and (ii) empirically: (i) . . . . (ii) . . . .

7 Write down the relationships between the angle of shearing resistance ϕ' and the coefficients of active and passive earth pressure respectively: (i) . . . .
(ii) . . . .

8 The Rankine/Bell equations giving respectively the active and passive earth pressures in cohesive soils are: p_a = ... and p_p = ...

9 Coulomb's . . . . may also be used to calculate lateral earth pressure and forces on retaining walls.

10 The principal forces tending to cause instability in a natural or man-made slope are . . . . and . . . .

ESSAY QUESTIONS

1 Define the terms **effective stress, pore pressure** and **total stress** and explain their significance in engineering design.

2 Explain the soil failure condition when in a state of plastic equilibrium. Why is this known as a **limit state**?

3 Distinguish between **active, passive** and **at-rest** stress states in the context of retaining wall design.

4 Compare Rankine's earth pressure theory with Coulomb's wedge theory, commentin on the relative advantages and disadvantages of each.

5 Discuss the factors affecting the stability of natural and man-made slopes, and explain why changes in certain conditions may bring about a slope failure.

CALCULATION EXERCISES (answers on page 209)

1 The soil layers on a site consist of a horizontal surface layer of sandy gravel 5 m thick, which overlies a layer of clay 4.5 m thick, which in turn is underlain by impermeable rock. Draw an effective-stress/total-stress profile between 0 m and 9.5 m assuming a water table level of 2.0 below the surface.

Unit weights:	sandy-gravel (saturated)	= 20.6 kN/m³
	sandy-gravel (drained)	= 19.0 kN/m³
	clay	= 18.0 kN/m³

2 A 3 m layer of graded fill is laid over a horizontal layer of sandy silt. Calculate the effective and total vertical stresses at a depth of 2 m below the top of the sandy silt when the water table is:
(a) at the top of the silt;
(b) at the top of the fill.

Unit weights: Fill (saturated)	= 21.5 kN/m³
(drained)	= 20.0 kN/m³
Sandy-silt (saturated)	= 19.4 kN/m³

3 The plan of a raft foundation is shown in *Fig 11.20*; it is to transmit a uniform contact pressure of 140 kN/m². Using the influence factors shown in *Table 11.1*, calculate the vertical direct stress occurring:
(a) at a point 10 m below A;
(b) at a point 5 m below B.

4 Evaluate the coefficients of earth pressure (K_o, K_a, K_p) for soils having angles of shearing resistance of:
(a) 25°; (b) 12°.

5 A retaining wall having a smooth vertical back is required to retain 9 m thickness of a homogeneous cohesionless soil, having a horizontal surface and the following properties:
$c' = 0$ $\quad \phi' = 28°$ $\quad \gamma = 19.0$ kN/m² $\quad \gamma_{sat} = 20.5$ kN/m²

Determine the active lateral thrust on the wall when:
(a) the soil is fully drained;
(b) the soil is fully waterlogged.

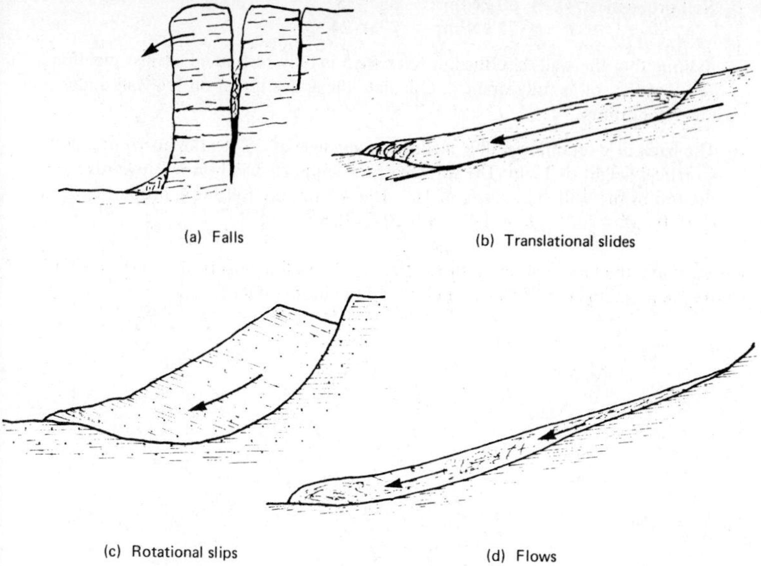

Fig 11.19 Modes of slope failure

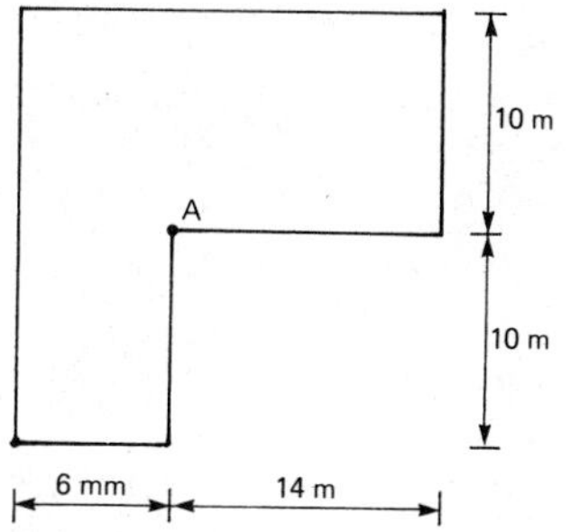

Fig 11.20

6 A concrete retaining wall having a smooth vertical back face is required to retain a 15 m thickness of homogeneous cohesionless soil below a horizontal surface. The soil has an angle of shearing resistance of 30°, a drained density of 19.5 kN/m^3 and a saturated density of 20.5 kN/m^3. Determine the total lateral thrust acting on the wall when the water table stands at a depth below the surface of 6 m.

7 Determine the total active thrust acting on the wall described in *Exercise 5* above when the soil is fully drained, but a uniform surcharge of 40 kN/m^2 is applied to the surface.

8 A concrete retaining wall with a smooth vertical back is to support a 15 m thickness of cohesive soil. Determine the active lateral thrust (a) in immediate undrained conditions and (b) in long-term fully-drained conditions.

Soil properties: $c_u = 60 \text{ kN/m}^2$ $\phi_u = 5°$ $\gamma = 18 \text{ kN/m}^3$
$c' = 22 \text{ kN/m}^2$ $\phi' = 24°$

9 Assume that the wall described in *Exercise 5* is fully restrained against yielding and that the soil is fully drained. Calculate the active thrust on the wall under these conditions.

10 The back of a retaining wall is inclined at an angle of 75° to the horizontal and has a vertical height of 12 m. The surface of the supported soil slopes upwards from the top of the wall at an angle of 16°. The soil has the following properties:
$c' = 0$ $\phi' = 32°$; $\delta = 24°$; $\gamma = 20.6 \text{ kN/m}^3$

Determine the lateral active thrust acting on the wall due to trial wedges having failure planes inclined at 50°, 60° and 70° to the horizontal.

12 Compressibility and settlement

12.1 MECHANISMS OF SETTLEMENT

Compression is a change of volume in a soil mass, while **settlement** is the vertical displacement of a structure due to interaction between the soil and the structure. Any mechanism that can bring about a volume change is a potential cause of settlement.

Structures vary in their response to settlement. Brittle construction (e.g. brick, masonry, unreinforced concrete) may sustain damage following very small displacements. More resilient construction (e.g. timber, steel or reinforced concrete) may be able to tolerate considerable displacements without undue distress.

Some of the common mechanisms which may lead to ground movement and settlement are summarised below:

COMPACTION

When mechanical energy is applied to a loose soil, its particles are forced closer together, air is expelled and if the water content remains the same, compaction is taking place. The energy required may result from self-weight, surface surcharge loading or from vibration (due to machinery, traffic, piling, etc.).

Loosely-packed sands and fill material are most susceptible.

CONSOLIDATION

Consolidation occurs when porewater is squeezed from the soil due to an increase in load; a decrease in load likewise may cause swelling. Saturated silts and clays are most susceptible.

(This process is explained more fully in section 12.3).

ELASTIC COMPRESSION

Soil grains are elastic bodies, so that some elastic distortion occurs immediately a load is applied; the resulting displacement is called **immediate settlement**.

(This process is explained more fully in section 12.2).

MOISTURE MOVEMENT

Certain clay soils shrink or expand considerably following changes in moisture content; they are referred to as **shrinkable clays**, or sometimes **expansive clays** (see also Section 9.1).

Seasonal climatic variation leads to changes in moisture content down to a depth of about 1 m, producing, for example in the London Clay, annual movements of 20–30 mm. It is recommended that foundations in these soils be placed at a depth of at least 1.25 m in order to avoid settlement.

VEGETATION TRANSPIRATION

Trees absorb water from the soil through their roots and expel it through their leaves. Certain types of fast-growing trees and also large trees draw in considerable quantities of water from the surrounding soil.

New plantings of seedlings in shrinkable soils should not be closer to the building than 1½ times their mature height. Removal of such trees in similar conditions leads to swelling and building damage may result due to uplift.

GROUNDWATER LOWERING

Water pumped from an excavation will lower the groundwater level in the surrounding soil. Two effects may result from this: (i) if the soil is shrinkable, a volume reduction will take place; (ii) a reduction in pore pressure will increase the effective overburden stress, which may then lead to consolidation settlement of the soil beneath the reduced groundwater level.

LARGE TEMPERATURE CHANGES

Shrinkage may take place in clay soils as they dry out under furnaces, kilns, boilers, etc. It is often necessary to provide a cooling air circulation beneath such constructions.

In silts, fine sands and chalky soils, sustained freezing conditions may produce **frost heave**, due to the expansion of groundwater as it freezes and the subsequent build up of **ice-lenses**. Frozen soil also has an abnormally high moisture content, so that immediately on **thawing** it may have a much reduced shear strength.

PIPING AND SCOURING

The phenomena of **quicksand** and **piping** are explained in sub-section 10.4. Settlement may occur in **quick** conditions due to the loss of shear strength.

Scouring is the removal of soil by moving water, such as in streams in flood, or on slopes, or where buried water pipes have been fractured.

LOSS OF LATERAL SUPPORT

The effectiveness of a foundation soil in providing support for a footing depends on it being confined by the depth of soil on either side. If this lateral support is removed, as for example when an adjacent excavation is dug, a shear slip may result carrying the footing into the excavation (*Fig 12.1*).

Similarly, settlement of a building at the top of a slope may be the result of slope movement.

SUBSIDENCE

Subsidence is a general term applied to settlement which is due to the collapse of an underground structure or cavity. Typical subsidence circumstances include collapse of abandoned mines, shafts, culverts, sewers, solution caverns and caves.

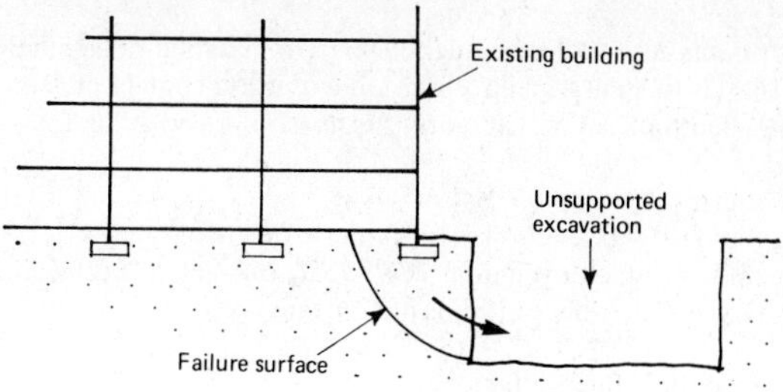

Fig 12.1 Loss of lateral support

12.2 IMMEDIATE SETTLEMENT

If the soil mass beneath a loaded area is assumed to behave as an elastic body, the vertical elastic displacement, or **immediate settlement**, (s_i) is given by the following expression:

$$s_i = \frac{qB}{E}(1 - \nu^2) I_\rho$$

where q = intensity of contact pressure;
B = breadth of the loaded area;
ν = Poisson's ratio (between 0.25 and 0.5);
E = Modulus of elasticity (determined from laboratory shear tests);
I_ρ = influence factor for vertical displacement (values of which are given in *Table 12.1*)

TABLE 12.1 Influence factors (I_ρ) for vertical displacement due to elastic compression

Shape	*Flexible*			*Rigid*
	Centre	*Corner*	*Average*	
Circle	1.00	0.64	0.85	0.79
Rectangle				
$\frac{L}{B}$ = 1.0	1.122	0.561	0.946	0.82
1.5	1.358	0.679	1.148	1.06
2.0	1.532	0.766	1.300	1.20
3.0	1.783	0.892	1.527	1.42
4.0	1.964	0.982	1.694	1.58
5.0	2.105	1.052	1.826	1.70
10.0	2.540	1.270	2.246	2.10
100.0	4.010	2.005	3.693	3.47

Worked example 12.1 A rectangular concrete foundation of length 30 m and breadth 16 m is to be designed to transmit a uniform contact pressure of 210 kN/m^2 at a depth of 2.4 m. Laboratory tests have provided the following soil properties:
Modulus of elasticity (E) = 55 MN/m^2
Poisson's Ratio (ν) = 0.5. Bulk density (γ) = 19 kN/m^3
Using the influence factor values in *Table 12.1*, calculate the immediate elastic settlement expected at the centre of the footing.

Firstly, determine the influence factor

$$\frac{L}{B} = \frac{30}{16} = 1.875$$

From *Table 12.1* (interpolating)

$$I_\rho = 1.358 + (1.532 - 1.358)\left(\frac{1.875 - 1.500}{2.000 - 1.500}\right)$$

$$= 1.489$$

Then immediate settlement,

$$s_i = \frac{qB}{E}(1 - \nu^2)\, I_\rho$$

$$= 210 \times 16(1 - 0.5^2)\, 1.489 \times 10^3 = \mathbf{68\ mm}$$

12.3 CONSOLIDATION SETTLEMENT

If a saturated cohesive soil is subject to an increase in applied stress, the pore pressure is immediately increased by the same amount. The excess pore pressure causes water to flow away at a rate controlled by the permeability of the soil, resulting in a reduction in volume. This process is called **consolidation.**

In 1943, Karl Terzaghi suggested a model to illustrate the mechanics of the process. Steel springs representing the soil fabric are enclosed in a cylinder full of water (*Fig 12.2a*). The increase in stress is applied through a piston, with the soil's permeability being simulated by a drainage valve.

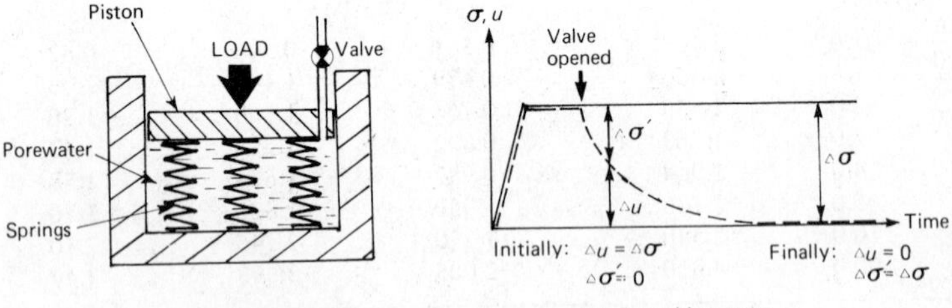

(a) Terzaghi's model for one-dimensional consolidation

(b) Stress/time curve

Fig 12.2 One-dimensional consolidation

If a load is applied to the piston with the valve closed, the length of the springs remains unchanged, since water is incompressible. If the applied load increases the total stress by $\Delta\sigma$, the pore pressure must increase by the same amount: $\Delta u = \Delta\sigma$. When the valve is opened, this excess pore pressure will cause water to flow out of the cylinder; the pore pressure decreases (rapidly at first, but slowing down) and the applied stress is then transferred to the springs (*Fig 12.2b*). At a given time, t:

the pore pressure, $u_t = u_o + \Delta u$
the effective stress, $\sigma'_t = \sigma'_o + \Delta\sigma'$, and
the increase in effective stress (causing the change in length of the springs) will be:
$\Delta\sigma' = \Delta\sigma - \Delta u$

The process continues until the whole of excess pore pressure has been dissipated and the whole of the increase in stress transferred to the springs (soil fabric). The **rate** of compress depends on how much the valve is opened, this being analogous to the permeability of the soil.

AMOUNT OF SETTLEMENT

If the soil is saturated, the change in water volume during consolidation will be equal to the change in void volume. *Fig 12.3* shows a model soil sample subject to an increase in

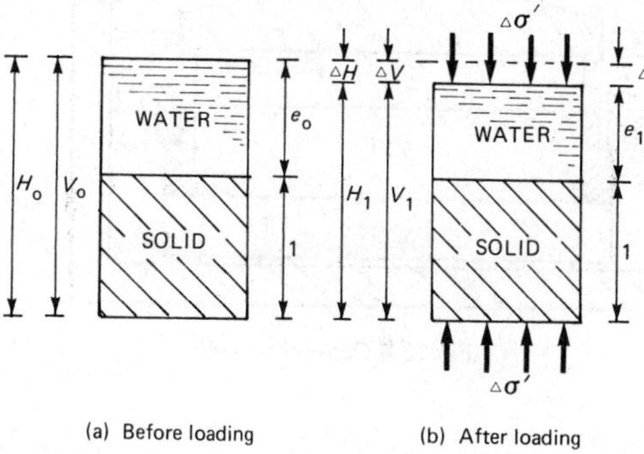

Fig 12.3 Compressibility explained using soil model

effective stress ($\Delta\sigma'$). If no **lateral** expansion takes place, the volumetric strains may be equated thus

$$\frac{\Delta V}{V_o} = \frac{\Delta H}{H_o} = \frac{\Delta e}{1 + e_o}$$

The change in thickness of a layer of soil initially H_o thick is therefore

$$\Delta H = \frac{\Delta e}{1 + e_o} H_o$$

However, the volumetric strain is also a function of the increase in stress, i.e.

$$\frac{\Delta e}{1 + e_o} = m_v \; \Delta\sigma'$$

where m_v = coefficient of volume compressibility (i.e. change in unit volume per unit change in effective stress)

Then **consolidation settlement,** $s_c = \Delta H = m_v \Delta\sigma' H_o$

The relationship between the void ratio and effective stress is measured in the one-dimensional consolidation test, or **oedometer test.**

12.4 OEDOMETER TEST

The apparatus, shown in *Fig 12.4*, consists of a cell (vessel) containing water and a device for applying a constant load. A disc of soil (usually 75 mm dia X 15–20 mm thick) is cut from an undisturbed sample and trimmed to fit inside a metal ring. The soil specimen, sandwiched between porous stones, is placed in the cell and held in place by a clamping ring.

A vertical static load is applied through a lever system and the change in thickness of the sample measured at intervals, using either a dial gauge or a displacement transducer.

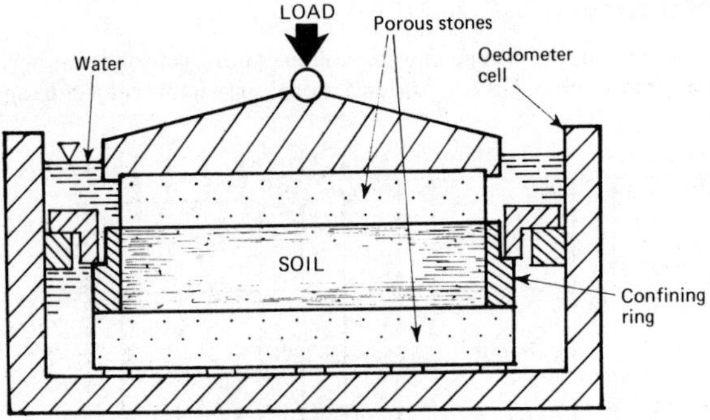

Fig 12.4 Oedometer cell

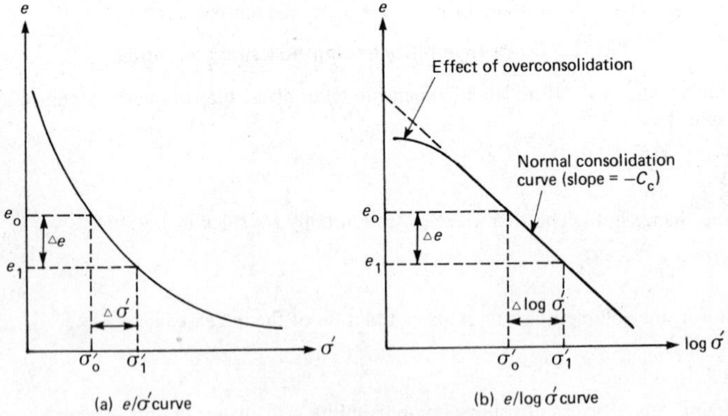

Fig 12.5 Compression curves

Readings are taken at various time intervals until the sample is fully consolidated, or for a set period of 24 or 48 hours. Further increments of load are added and the readings continued for the same time interval after each; each load increment is usually double the previous one. The number and value of the load increments is chosen in accordance with the type of soil and the range of stress expected.

After the sample has been allowed to consolidate under the final load, the whole load is removed and the sample allowed to swell for (usually) 24 hours. After this, the soil specimen is removed from the apparatus and its moisture content (m_f) recorded. This enables the final void ratio (e_f) to be determined:

$$e_f = m_f G_s \text{ (since } S_r = 1 \text{ for a saturated soil)}$$

From the observed changes in thickness (Δh), corresponding changes in void ratio for each stage can be evaluated:

$$\Delta e = \frac{\Delta h}{h_o}(1 + e)$$

and void ratio/effective stress graphs plotted as shown in *Fig 12.5*.

12.5 CALCULATION OF CONSOLIDATION SETTLEMENT

If, in a problem, the effective stress changes from σ'_0 to σ'_1 the corresponding void ratios e_o and e_1 can be obtained from the e/σ' graph (*Fig 12.5a*).

Then consolidation settlement, $$s_c = \frac{\Delta e}{1 + e_o} H_o$$

$$= \frac{e_o - e_1}{1 + e_o} H_o$$

where H_o = thickness of soil layer.

Alternatively values can be determined for m_v, since

$$m_v = \frac{\Delta e}{\Delta \sigma'} \cdot \frac{1}{1 + e_o}$$

$$= \frac{e_o - e_1}{\sigma_2 - \sigma_o} \cdot \frac{1}{1 + e_o}$$

For **normally-consolidated*** soils a more reliable interpretation of the oedometer results may be obtained by plotting $e/\log \sigma'$ curves (*Fig 12.5b*), for which a straight-line fit may be used. The slope of the straight portion is referred to as the **compression index** (C_c)

$$C_c = -\frac{\Delta e}{\Delta \log \sigma'} = \frac{e_o - e_1}{\log (\sigma'_1/\sigma_o')}$$

The change in void ratio is then

$$e_o - e_1 = C_c \log (\sigma'_1/\sigma'_o)$$

and the consolidation settlement

$$s_c = \frac{C_c}{1 + e_o} \log (\sigma_1'/\sigma'_o) H_o$$

*A **normally-consolidated** clay has not been subject to a stress higher than its present-day overburden stress, whereas an **over-consolidated** clay has (in the past) been subject to greater stress. Overconsolidated clays (e.g. London Clay) tend to dilate (expand) when subject to shear or when exposed to weather.

12.6 CALCULATION OF SETTLEMENT TIME

For design purposes, it is equally as important to calculate the time required for settlement to take place as it is to determine the amount. The length of time required for a given fraction of the total settlement may be obtained from the following expression

$$t = \frac{T_v d^2}{c_v}$$

where T_v = a 'time factor', dependent on both the degree (fraction) of consolidation and the distribution of stress through the thickness of the layer, usually obtained from a chart or table (not given here);

d = length of drainage path = h or ½h (see *Fig 12.6*);

c_v = coefficient of consolidation.

One method of obtaining a value for c_v corresponding to a particular stress range fror the oedometer test results is described below.

Choosing a loading increment stage appropriate to the stress range required, the chang in thickness are plotted against the square root of time (*Fig 12.7*) and a straight line

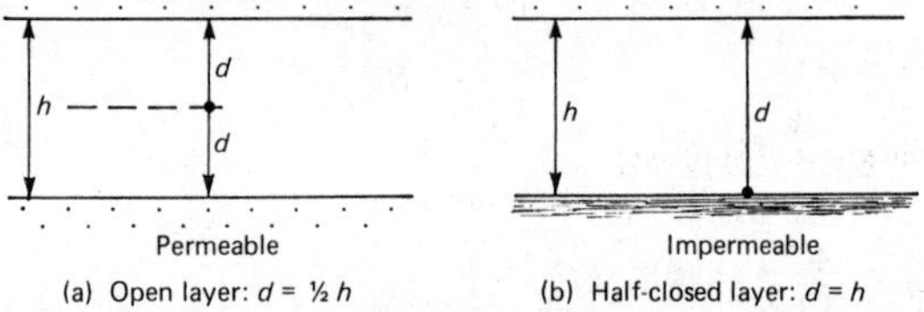

Fig 12.6 Drainage paths

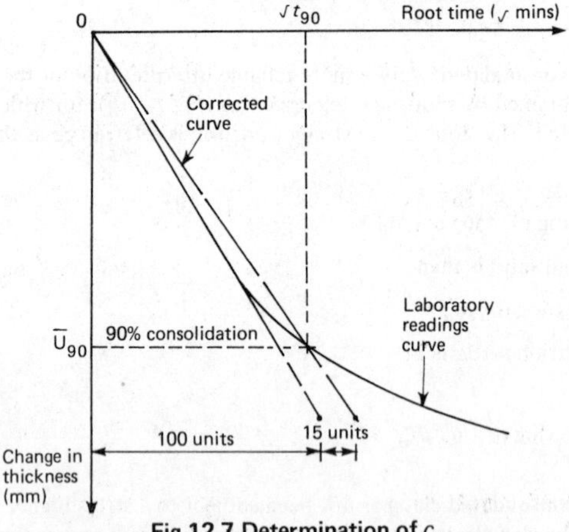

Fig 12.7 Determination of c_v

drawn through the first 50–60% of the curve. A second straight line is now drawn having abscissae 1.15 times longer than the first. The 'corrected' straight line is assumed to cut the experimental curve at the 90% consolidation point ($\bar{U} = 0.9$), giving an intercept on the time axis of $\sqrt{t_{90}}$

Then coefficient of consolidation, $c_v = \dfrac{T_{v(90)} d^2}{t_{90}}$

For the oedometer test, $T_{v(90)} = 0.848$

$$\therefore c_v = \frac{0.848\, d^2}{t_{90}}$$

For a given soil: $\dfrac{T_v}{c_v} = \dfrac{t}{d^2} = \text{constant}$

Then to relate laboratory results with site times:

$$\frac{t_{lab}}{d^2_{lab}} = \frac{t_{site}}{d^2_{site}}$$

Worked example 12.2 The data given below was recorded during an oedometer test on a specimen of a saturated clay. Each load increment was held constant for 24 hr before adding the next. At the end of the last load period, the load was removed entirely and the sample allowed to expand for 24 hours.

Final thickness after expansion = 16.70 mm
Final moisture content after expansion = 30.6%
Specific gravity = 2.66.

Applied stress (kN/m^2)	0	25	50	100	200	400	800
Thickness (mm)	18.80	18.39	18.17	17.71	17.23	16.73	16.22

(a) Plot the e/σ' curve and determine the coefficient of volume compressibility (m_v) corresponding to a change in effective stress from 188 kN/m^2 to 350 kN/m^2.
(b) Plot the $e/\log\sigma'$ curve and determine the compressibility index (C_c).
(c) Plot the m_v/σ' curve.
(d) Obtain an estimate for the consolidation settlement to be expected when a 4 m thick layer of this soil is subject to an average change in effective stress from 180 kN/m^2 to 350 kN/m^2.

Firstly, the final void ratio is determined.
Since the specimen is saturated, $S_r = 1.0$
Therefore $e = mG_s = 0.306 \times 2.66 = 0.814$

Change in void ratio, $\Delta e = \dfrac{\Delta h}{h}(1 + e)$

During the swelling stage, $\Delta e = \dfrac{0.75}{16.70}(1.814) = 0.082$

During the 400–800 stage, $\Delta e = \dfrac{-0.48}{16.25}(1.732) = -0.051$

Thus the void ratios corresponding to the beginning of each stage may be computed. The full set of calculations is tabulated below.

σ' (kN/m²)	$\Delta\sigma'$ (kN/m²)	h (mm)	Δh (mm)	Δe	e	$\log \sigma'$	$\dfrac{\Delta e}{\Delta\sigma'}$	m_v (10^{-3} m²/kN)
0		18.80			1.003			
	25		−0.41	−0.044			1.76	0.82
25		18.39			0.959	1.40		
	25		−0.22	−0.023			0.92	0.45
50		18.17			0.936	1.70		
	50		−0.46	−0.049			0.98	0.51
100		17.71			0.887	2.00		
	100		−0.48	−0.051			0.51	0.27
200		17.23			0.836	2.30		
	200		−0.50	−0.053			0.265	0.14
400		16.73			0.783	2.60		
	400		−0.48	−0.051			0.128	0.067
800		16.25			0.732	2.90		
			+0.75	0.082				
0		17.00			0.814			

(a) *The e/σ' curve plot is shown in Fig 12.8*

Reading off the curve: for $\sigma'_o = 180$ kN/m², $e_o = 0.844$
for $\sigma'_1 = 350$ kN/m², $e_1 = 0.793$

Coefficient of volume compressibility, $m_v = \dfrac{\Delta e}{\Delta\sigma'} \cdot \dfrac{1}{1 + e_o}$

$$= \frac{0.844 - 0.793}{(350-180)1.846}$$

$$= \mathbf{0.163 \times 10^{-3}\ m^2/kN}$$

(b) *The $e/\log\sigma'$ curve plot is shown in Fig 12.9*
The slope of the straight portion of this curve gives the compressibility index (C_c)

$$C_c = \frac{\Delta e}{\Delta \log\sigma'}$$

$$= \frac{0.887 - 0.732}{2.903 - 2.000} = 0.172$$

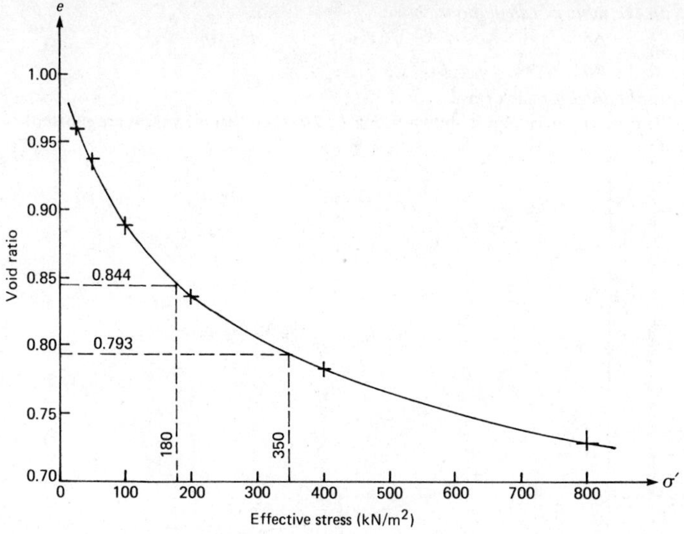

Fig 12.8

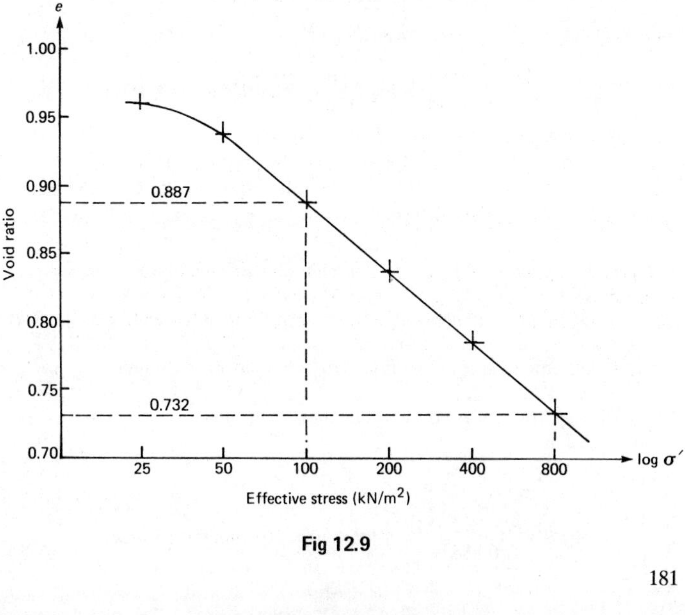

Fig 12.9

(c) *In the table of calculations above,*

$$m_v = \frac{\Delta e}{\Delta \sigma'} \cdot \frac{1}{1 + e_o}$$

is obtained for each stage.

The m_v/σ' curve plot is shown in *Fig 12.10* (end-of-stage values are plotted).

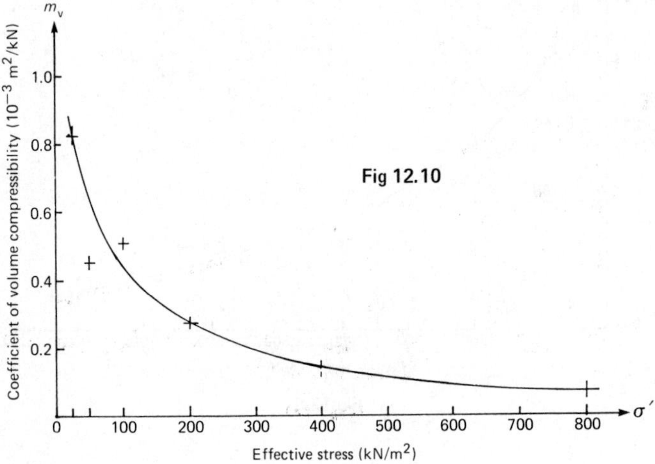

Fig 12.10

(d) From part (a): $s_c = m_v \, \Delta\sigma' \, H_o$

$$= 0.163 \times 10^{-3}(350 - 180) \times 4 \times 10^3 = \mathbf{111\ mm}$$

From part (b): $s_c = \dfrac{C_c}{1 + e_o} \log (\sigma'_1/\sigma'_o) \, H_o$

$$= \frac{0.172}{1.844} \times \log \frac{350}{180} \times 4 \times 10^3 = \mathbf{108\ mm}$$

From part (c): the m_v/σ' curve yields a value of

$$m_v = 0.16 \times 10^{-3} \text{ m}^2/\text{kN (at } \sigma' = 350 \text{ kN/m}^2)$$

Then $s_c = 0.160 \times 10^{-3} (350 - 180) \times 4 \times 10^3 = \mathbf{109\ mm}$

Worked example 12.3 The final consolidation settlement estimated for 4.4 m thick layer of clay is 146 mm. The coefficient of consolidation is 0.855 m²/minute and permeable layers occur both above and below the clay. Calculate the time required for:
(a) 90% of the final consolidation; (b) a settlement of 50 mm.

(a) For two-way drainage at 90% consolidation $T_v = 0.848$ and $d = 2.2$ m

Then $t_{90} = \dfrac{T_{v(90)} d^2}{c_v}$

$$= \frac{0.848 \times (2.2 \times 10^3)^2}{0.855} = 4.80 \times 10^6 \text{ min} = \mathbf{9.13\ years}$$

(b) A settlement of 50 mm represent a degree of consolidation of

$$\bar{U} = \frac{50}{146} = 0.342\ (34.2\%)$$

The value of T_v corresponding to this degree of consolidation has to be obtained from an appropriate table or curve (not included in this book).

$T_{v(34.2)} = 0.094$

Then for a settlement of 50 mm:

$$t = \frac{0.094\,(2.2 \times 10^3)^2}{0.855}$$

$= 0.532 \times 10^6$ min $=$ **1.07 years.**

Note: For a given soil the coefficient of consolidation (c_v) is not usually constant, so that the above calculations yield **estimates** only.

Worked example 12.4 In a laboratory oedometer test the time required during one of the loading stages for 50% of the total compression was 16.2 min, at which point the sample thickness was 18.64 mm. Determine the time required for 50% of the final consolidation settlement to take place in a 5 m thick layer of the same soil when:
(a) drained top and bottom; (b) drained from one surface only.

(a) In the laboratory specimen the drainage path, $d_{lab} = \frac{18.64}{2}$ mm

On site the drainage path, $d_{site} = \frac{5 \times 10^3}{2}$ mm

From $\frac{t_{lab}}{d^2_{lab}} = \frac{t_{site}}{d^2_{site}}$, $\quad t_{site} = \frac{16.2 \times (5 \times 10^3)^2}{18.64^2}$

$= 1.166 \times 10^6$ min

$=$ **2.22 years**

(b) In the laboratory specimen the drainage path is still $d_{lab} = \frac{18.64}{2} = 9.32$ mm

On site, however, the drainage path, $d_{site} = 5 \times 10^3$ mm

In this case, $\quad t_{site} = \frac{16.2 \times (5 \times 10^3)^2}{9.32^2}$

$= 4.66 \times 10^6$ min

$=$ **8.87 years**

TEST EXERCISES

COMPLETION QUESTIONS (answers on page 209)

Complete the following statements by inserting the most appropriate word or words in the spaces indicated.

1 State **five** common mechanisms which may lead to ground movement and settlement . . . .
. . . .

2 .... compression takes place when the displacements are entirely proportional to the intensity of loading.

3 The process whereby **air** is expelled during volume change is called ....

4 The process of volume change resulting from the squeezing out of pore water is called ....

5 The following volumetric strain relationship is used in the determination of consolidation settlement,

$$\frac{\Delta V}{V_o} = \frac{\Delta H}{\ldots} = \frac{\Delta e}{\ldots}$$

6 If m_v = coefficient of volume compressibility, $\Delta\sigma'$ = change in effective stress, then the consolidation settlement of a layer of thickness H_o is given by:
$s_c = \Delta H = \ldots\ldots\ldots$

7 The .... test is a one-dimensional consolidation test normally used to determine the settlement characteristics of a soil.

8 If consolidation test results are plotted for a .... soil in the form of a $e/\log\sigma'$ graph, a straight line may be fitted through the points.

9 The slope of the straight-line portion of the $e/\log\sigma'$ graph is termed the .... index of the soil.

10 The time required for a given degree of consolidation settlement to take place is given by
$t = \ldots\ldots\ldots$ where T_v = time factor;
d = drainage path length;
c_v = coefficient of consolidation.

11 An .... clay has been in the past subjected to a greater stress than its present-day overburden stress.

12 The London Clay is an example of an .... clay.

13 Such clays (as in 11 and 12) tend to .... when subject to shear or when exposed to weather.

ESSAY QUESTIONS

1 Discuss the response of various types of structure to ground movement. What constitutes a critical condition(s) and how does the nature of the structure itself affect this?

2 List and discuss the various mechanisms that may lead to foundation settlement. Indicate where appropriate the type of soil or any other critical factor controlling the mechanism.

3 Describe a model soil subject to compression and use it to explain the processes (a) elastic compression and (b) consolidation.

4 Distinguish between the terms **normally-consolidated** and **overconsolidated** when applied to clay soils. What are the significant behavioural characteristics of each type when subject to shearing?

CALCULATION EXERCISES (answers on page 209)

1 A rectangular raft foundation of length 12 m and 5 m is to be founded at a depth of 1.8 m in a saturated clay soil having an undrained modulus of elasticity of 68 MN/m^2. Determine the amount of immediate (elastic) settlement likely to occur at the centre of the foundation due to a uniform applied pressure of 180 kN/m^2.

2 The following data was recorded during an oedometer consolidation test on a specimen of a saturated clay. Each load stage was maintained for 24 hours and a final 24 hour swelling stage concluded the test.

Applied stress (kN/m^2)	0	50	100	200	400	800
Thickness (mm)	19.59	19.18	19.03	18.76	18.47	18.19

At the completion of the swelling stage the specimen had a thickness of 18.80 mm and a moisture content of 29.5%.

(a) Assuming a specific gravity of 2.68, calculate the void ratios at the end of each load stage and plot the void ratio/log effective stress curve.

(b) From the e/log σ' curve, determine the compression index and hence an estimate of the total settlement to be expected due to the consolidation of a 5 m thick layer of this soil when the average effective stress in the layer is increased from 125 kN/m^2 to 310 kN/m^2.

3 A layer of saturated clay soil having a thickness of 4.8 m is subjected to an increase in effective stress of 220 kN/m^2. Calculate the final settlement due to the consolidation of the layer if the coefficient of volume compressibility is 0.32 m^2/MN.

4 A clay layer of thickness 3.8 m is underlain and overlain by permeable strata. The construction loading will induce a uniform increase in effective stress throughout the clay. In an oedometer test the same change in effective stress produced a total change in void ratio from 0.796 to 0.768; after 15.5 min 50% of this change had taken place, at which point the sample thickness had been reduced to 18.26 mm.

(a) Calculate the total consolidation settlement expected.

(b) Calculate the time required for 50% of the total settlement to take place.

5 It has been calculated that the average increase in effective stress throughout a 5.0 m thick layer of clay is 208 kN/m^2. From laboratory tests the following properties have been established for the soil.

Coefficient of compressibility, = 0.28 m^2/MN
Coefficient of consolidation, = 0.75 mm^2/min.

Calculate the amount of consolidation settlement to be expected corresponding to a degree of consolidation of 90%, and also the time required for this to take place (T_v = 0.848).

13 Shear strength of soil

13.1 PRINCIPLES AND PARAMETERS

The **shear strength** of a soil is defined as the maximum intensity of shear stress that may be induced within its mass, or as the intensity of shear stress along the failure surface at the time of failure.

Shear strength is therefore a limiting value corresponding to a state of plastic equilibrium. There are two components of shear strength: **friction** and **cohesion.**

INTERNAL FRICTION

Friction is developed between soil particles in contact with each other. The frictional resistance (T) is dependent on the normal force (N) at the surface of contact (*Fig 13.1a*). The limiting value of T increases proportionally with N; or in terms of stresses, the **shear strength** (τ_f) increases proportionally with the **normal stress** (σ_n) on the failure plane (RS).

$$\tau_f = \frac{T}{\text{contact area}}$$

$$\sigma_n = \frac{N}{\text{contact area}}$$

The graph drawn of the limiting values of τ_f against corresponding values of σ_n, is known as the **failure envelope** (*Fig 13.1b*). The slope of the failure envelope is

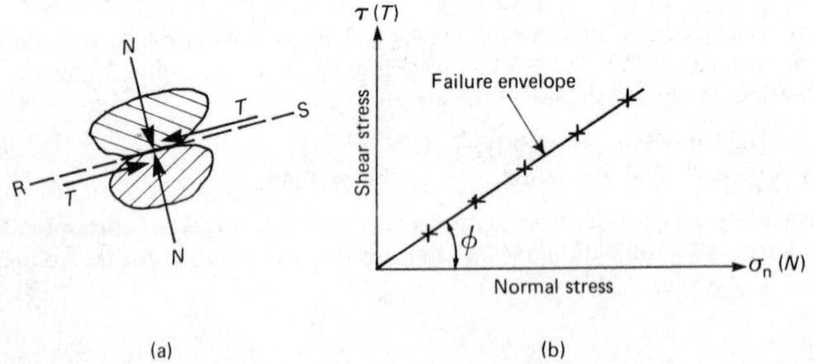

Fig 13.1 Frictional shear strength

referred to as the **angle of shearing resistance** (ϕ). Values of ϕ vary from zero for saturated clays and silts, up to 45° for coarse-grained angular gravels.

APPARENT COHESION

Cohesive soils such as silts and clays, exhibit a shear strength, even when the normal stress is zero; this is termed the **apparent cohesion** (c) of the soil. The value of c will be zero for clean sands and gravels, and will vary with the plasticity and moisture content of silts and clays.

COULOMB'S EQUATION

The values of c and ϕ are referred to as the **shear strength parameters** of the soil. An expression combining these to give a straight-line failure envelope was proposed by Coulomb in 1773; this is known as Coulomb's equation. (*Fig 13.2*)

$$\tau_f = c + \sigma_n \tan \phi$$

where c = apparent cohesion;
ϕ = angle of shearing resistance;
σ_n = normal stress on the failure plane.

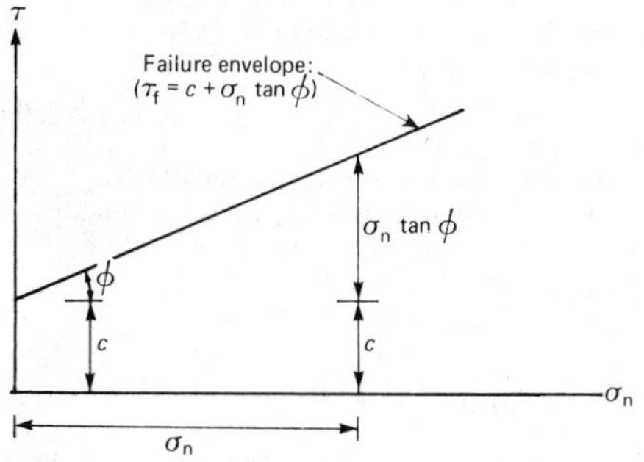

Fig 13.2 Coulomb's equation for the failure envelope

13.2 SOIL TYPES AND PRACTICAL PARAMETERS

In some soils, both cohesion and friction contribute to shear strength, in others only one component may be present. The classification of soils in terms of their shear strength characteristics is given below and illustrated in *Fig 13.3*.

Cohesionless soils such as sands and gravels, where $c = 0$, (*Fig 13.5a*).
Frictionless soils such as undrained saturated clays and silts, where $\phi = 0$ (*Fig 13.5b*).
$c - \phi$ soils partially saturated mixtures, such as clayey sand, silty sand, etc., where $c > 0$ and $\phi > 0$ (*Fig 13.2*).

[*Note*: when both $c = 0$ and $\phi = 0$, there is no shear strength, and the soil will behave as a liquid, e.g. muds, slurries.]

The values of c and ϕ used in design will have been obtained from shear tests (Section 13.4 and 13.5). It is important to note that the test conditions, particularly

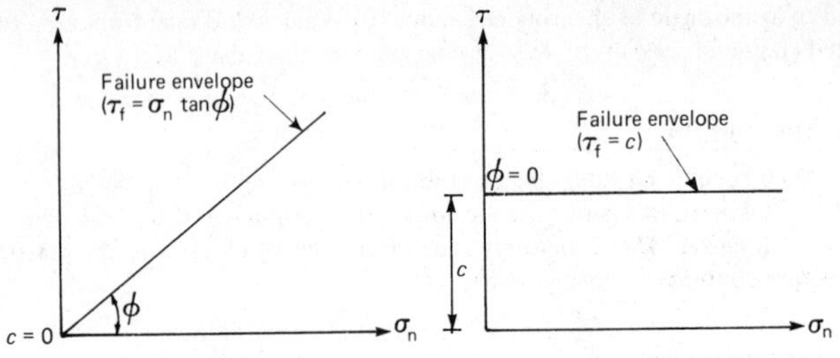

Fig 13.3 (a) Cohesionless soil (b) Frictionless soil

those relating to drainage, influence the measured values of both parameters. It is therefore necessary to indicate the test conditions, by adding a subscript to the approp symbol as follows:

c_u = apparent cohesion measured under undrained conditions;
ϕ_u = angle of shearing resistance measured under undrained conditions (in a fully-saturated soil, $\phi_u = 0$);
c' = apparent cohesion referred to effective stresses;
ϕ' = angle of shearing resistance referred to effective stresses. Coulomb's equatior referred to effective stresses is therefore:

$$\tau_f = c' + \sigma'_n \tan \phi' = c' + (\sigma_n - u_f) \tan \phi' \qquad (u_f = \text{pore pressure at failure})$$

c'_d = apparent cohesion under fully-drained conditions;
ϕ'_d = angle of shearing resistance under fully-drained conditions;

When the soil is fully drained, $u_f = 0$
Then $\tau_f = c'_d + \sigma_n \tan \phi'_d$

13.3 MOHR-COULOMB FAILURE THEORY

The forces acting at a given point in a soil may be resolved into three **principal stresses** acting at right angles to each other. A **principal** stress is defined as the normal stress acting on a **principal plane**, and a principal plane is a plane upon which no shear stress acts. The relationships between the principal stresses and the normal and shear stresses on other planes passing through the same point may be found using the **Mohr circle** construction.

Consider an element in a soil mass subject to principal stresses σ_1 and $\sigma_3 = \sigma_2$ (*Fig 13.4a*). It can be shown mathematically that on the plane PQ the variation in shear stress (τ) and normal stress (σ_n) is given by

$$\tau = \tfrac{1}{2}(\sigma_1 - \sigma_3) \sin 2\alpha$$
$$\sigma_n = \tfrac{1}{2}(\sigma_1 + \sigma_3) + \tfrac{1}{2}(\sigma_1 - \sigma_3) \cos 2\alpha$$

From which it may be seen that the plot of τ against σ_n is a circular graph (ADB), i.e. Mohr circle (*Fig 13.4b*).

Now suppose PQ is a failure plane, upon which the shear stress at failure (i.e. the **shear strength** of the soil) is τ_f. Point D therefore also lies on the Coulomb failure envelope indicated in *Fig 13.2*, and can be located by drawing the failure envelope so that it just touches the Mohr circle (*Fig 13.4*).

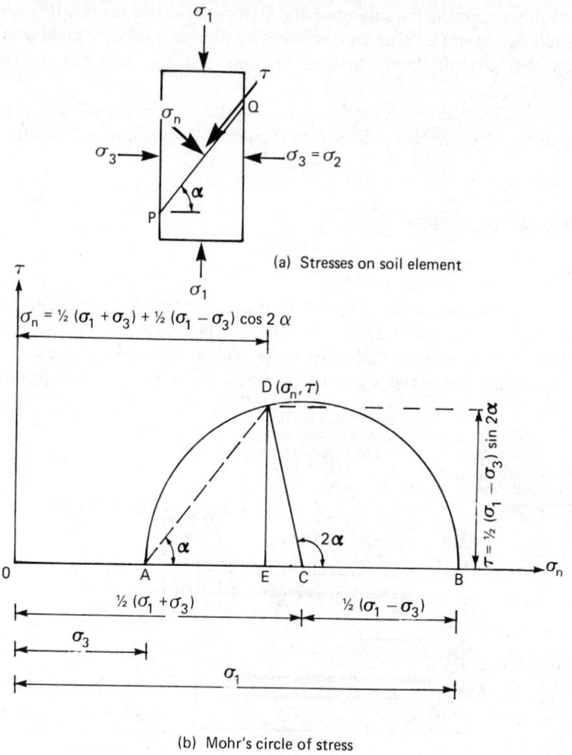

Fig 13.4

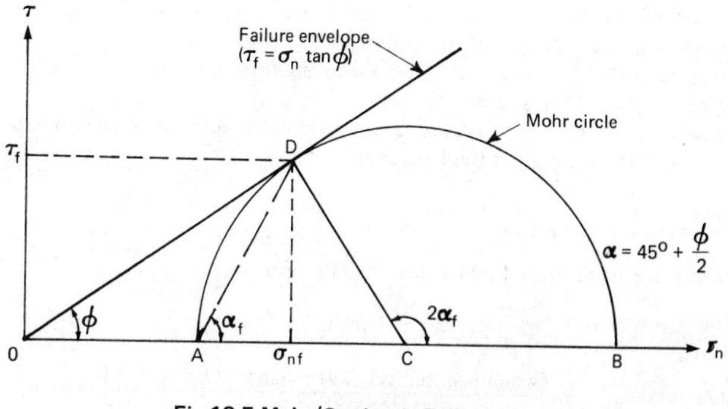

Fig 13.5 Mohr/Coulomb Failure theory

In order to determine the parameters c and $\emptyset$ from shear test results, it is necessary to draw the failure envelope. This may be done by directly measuring and plotting τ_f and σ_n obtained from a number of tests on the same soil, e.g. shear box test (section 13.4).

However, in the triaxial test (section 13.5), it is the principal stresses that are measured, and the Mohr circle construction then used to establish the failure envelope.

13.4 THE SHEAR BOX TEST

PROCEDURE

A rectangular prism of soil is carefully cut from an undisturbed sample and fitted into a square metal box, usually 60 mm × 60 mm (*Fig 13.6*). The box consists of upper and lower halves initially bolted together. The soil specimen, sandwiched between perforated metal plates and porous stones, and with a pressure pad on top,

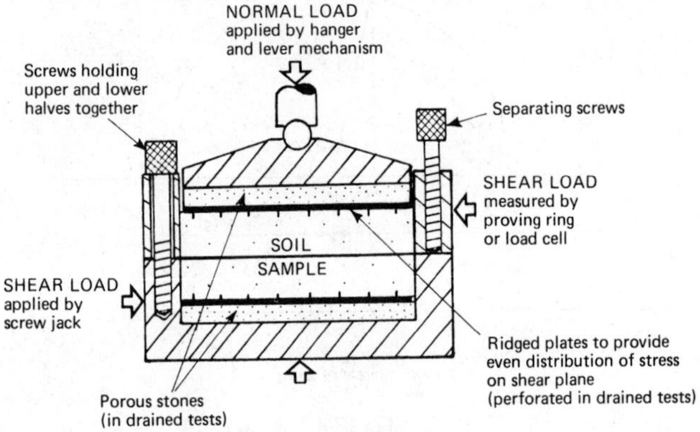

Fig 13.6 Shear box

is placed in the trolley and a vertical load applied by means of a weight-hanger assembly. After removing the screws holding the two halves of the box together, the soil specimen is sheared along a horizontal plane by applying a horizontal force through a screw jack. The magnitude of the shearing force is measured using either a proving ring or a load transducer.

This procedure is repeated several times on different specimens of the same soil, using different vertical loads on the hanger assembly.

INTERPRETATION OF RESULTS

1 Stresses on the shear plane (area = 0.06×0.06 m^2):

$$\text{shearing stress, } \tau = \frac{\text{shear load (kN)}}{0.06^2} \text{ kN/m}^2$$

$$\text{normal stress, } \sigma_n = \frac{(\text{weight on hanger (kg)} + \text{hanger (kg)})}{0.06^2} \times \frac{9.81}{10^3} \text{ kN/m}^2$$

2 For each specimen tested, a plot is drawn of shear stress against displacement, from which the maximum shear stress (τ_f) is determined.

3 Values of τ_f are now plotted against σ_n and the best straight line drawn through the points taken as the failure envelope.

4 Apparent cohesion (c) = intercept of failure envelope on the τ axis. Angle of shearing resistance ($\emptyset$) = angle of failure envelope to the σ_n axis.

Worked example 13.1 The following results were recorded during a shear box test on a cohesive soil.

Normal load (N)	73	191	309	427	545
Shear load at failure (N)	109	139	170	197	227

If the specimen size was 60 mm × 60 mm, plot the failure envelope and determine the apparent cohesion and angle of shearing resistance.

Area of shear plane = $60 \times 60 \times 10^{-6} = 3.6 \times 10^{-3}\ \text{m}^2$

The normal and shear stresses acting on the shear plane are obtained by dividing the appropriate load value by this area. For example, in the case of the first specimen, the normal load was 73 N and the shear load at failure was 109 N, therefore

Normal stress, $$\sigma'_n = \frac{73 \times 10^{-3}}{3.6 \times 10^{-3}} = 20.3\ \text{kN/m}^2$$

Shear stress at failure, $$\tau_f = \frac{109 \times 10^{-3}}{3.6 \times 10^{-3}} = 30.3\ \text{kN/m}^2$$

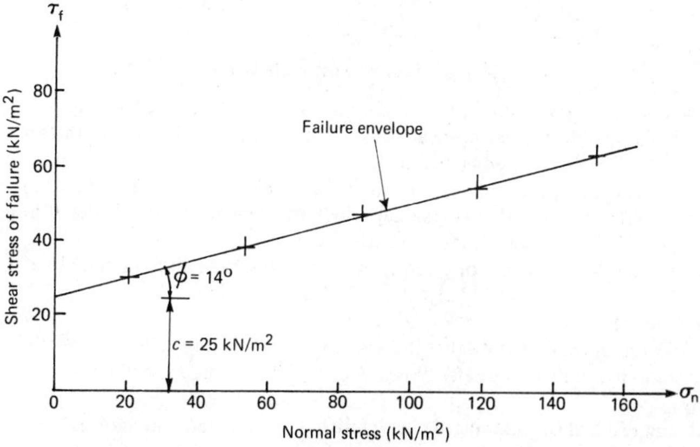

Fig 13.7

Similarly the other stress values will be:

Normal stress, σ'_n (kN/m²)	20.3	53.1	85.8	118.6	151.4
Shear stress at failure, τ_f (kN/m²)	30.3	38.6	47.2	54.7	63.1

The failure envelope is obtained by plotting τ_f against σ'_n and drawing the best straight line graph through the points. (*Fig 13.7*). The values of c' and ϕ' may then be read off th graph.

Apparent cohesion, $c' = 25$ kN/m²
Angle of shearing resistance, $\phi' = 14°$

PEAK AND RESIDUAL SHEAR STRENGTH

In densely compacted granular soils, such as sands, a high degree of interlocking is set up between particles. In order for a shear plane to develop the soil must dilate sufficiently for the interlocking to be overcome.

Initially, therefore, the shear stress rises sharply to a high **peak** value, at which the shear plane first begins to develop. As the shear strain increases further dilation occurs ar the shear stress falls back to an **ultimate** or **residual** value (*Fig 13.8*).

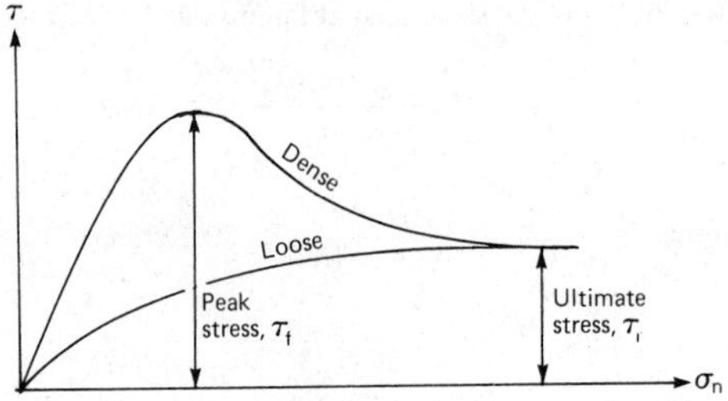

Fig 13.8 Peak and ultimate stress

In the case of a loosely-packed sand, compression takes place as shearing commences, accompanied by a slower rise in shear stress. As the shear strain increases the shear stress flattens out to reach the **residual** value.

The initial shape of the stress/strain curve is dependent on the state of packing and the difference between the peak and residual values also depends on the degree of interlocking made possible by the shape of particles.

The residual shear strength of a soil corresponds to shear failure taking place at consta volume; the soil is then said to be at its **critical state**, at which the void ratio is termed th **critical void ratio.**

It is most important to note that if a soil is sheared when in a denser state than its critical state, it will dilate initially. When a loose soil is sheared it will be compressed initially.

Peak and residual shear strength characteristics are illustrated in *Worked example 13.*

Worked example 13.2 Two specimens of a sand were tested in a shear box at the same constant normal stress of 200 kN/m^2. Specimen A was prepared in a loose state, while specimen B was compacted into a dense state.

The data given below was recorded during the tests.

(a) Plot the stress/strain and vertical displacement curves for the two tests.

(b) Determine the angle of shearing resistance corresponding to the loose and dense states.

TEST A : LOOSE STATE

Horizontal Displacement (10^{-2} mm)	0	50	100	150	200	250	300	350	400	450
Vertical (+*ve* = up) Displacement (10^{-2} mm)	0	−6	−9	−11	−13	−14	−14	−14	−14	−14
Shear Stress (kN/m^2)	0	35	57	72	79	84	88	89	90	90

TEST B : DENSE STATE

Horizontal Displacement (10^{-2} mm)	0	50	100	150	200	250	300	350	400	450
Vertical (+*ve* = up) Displacement (10^{-2} mm)	0	−3	+3	11	18	23	26	27	28	28
Shear Stress (kN/m^2)	0	67	111	124	120	103	94	88	88	87

(a) The plotted data is shown in *Fig 13.9*. The two upper curves show the variation in shear stress (τ) with horizontal (shear) displacement (Δl). In Test A, on the loose sand, τ rises slowly and steadily to reach an **ultimate** value of 88 kN/m^2.

In Test B, on the dense sand, the shear stress initially rises steeply reaching a **peak** value of 125 kN/m^2, before falling slowly to level off at the same approximate ultimate value of 88 kN/m^2.

The plots of change in thickness (Δh) against horizontal displacement show that, as the loose specimen is sheared, a reduction in volume takes place. As the dense specimen is sheared, however, after a very small compression at the start, a steady dilation (expansion) occurs.

(b) The 'dense' value of ϕ' is obtained by plotting the peak shear stress recorded for the dense specimen (125 kN/m^2) against the normal stress maintained during the test (200 kN/m^2). The failure envelope passes through this point and also the origin (*Fig 13.10*), giving $\phi'_{dense} = 32°$

The 'loose' value of ϕ' is obtained similarly by plotting the ultimate or residual value of τ_f obtained from Test A, i.e. 88 kN/m^2. Thus $\phi'_{loose} = 24°$.

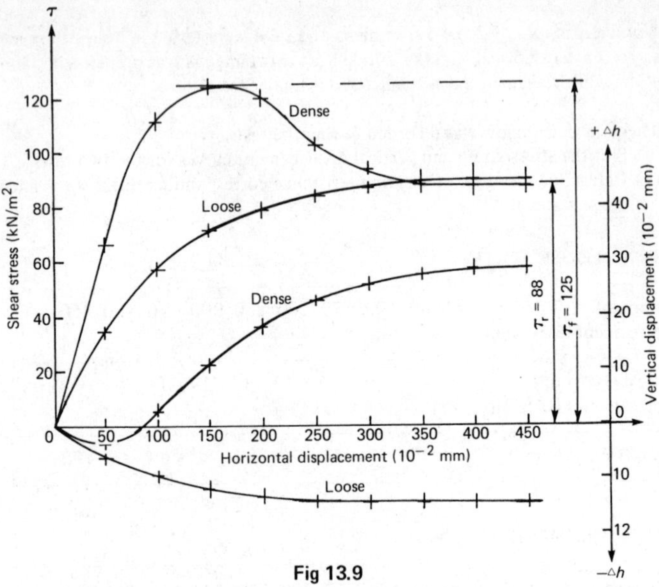

Fig 13.9

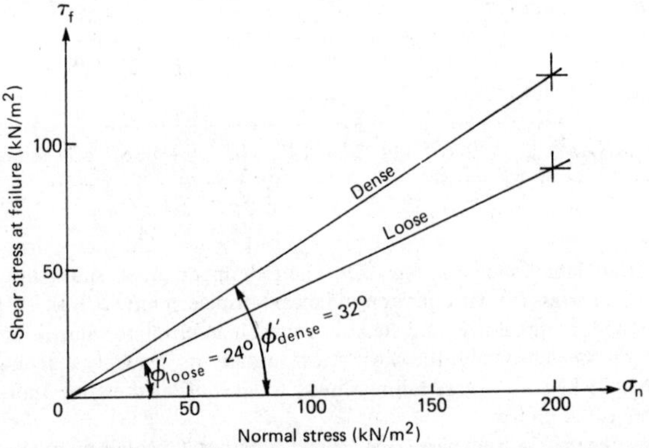

Fig 13.10

13.5 THE TRIAXIAL COMPRESSION TEST

PROCEDURE

The most widely used shear test is the triaxial compression test; it is suitable for all types of soil and allows a variety of conditions and procedures to be used. Three cylindrical specimens (usually 76 mm long × 38 mm diameter) are cut at the same depth level from an undisturbed soil sample. Each specimen is enclosed between

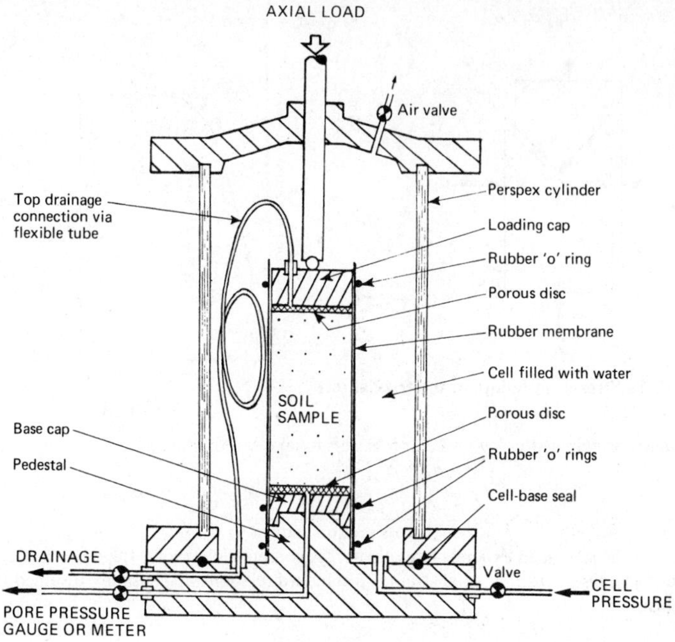

Fig 13.11 Triaxial cell

rigid end-caps inside a thin rubber membrane, with rubber o-rings to provide a water seal.

The prepared specimen is placed in the triaxial cell assembly as shown in *Fig 13.11*, with the rubber membrane stretched over the pedestal. When the cell has been re-assembled, it is filled with water and the air bled out. With the cell (water) pressure maintained at a prescribed constant value, the axial load on the specimen is increased at a constant rate of strain by means of a screw jack. The test is continued until either the specimen shears or a maximum vertical stress is reached.

READINGS

During the test the following readings are taken:

Vertical displacement: using either a dial gauge or a displacement transducer.

Axial load: using either a proving ring or a load transducer.

Pore pressure: using either a pressure gauge or a pressure transducer.

In tests of a specialised nature, certain other readings may be taken also, e.g. lateral displacement, volume change.

INTERPRETATION (see *Fig 13.12*)

1 Minor principal stress, $\sigma_3 = \sigma_2$ = cell pressure.

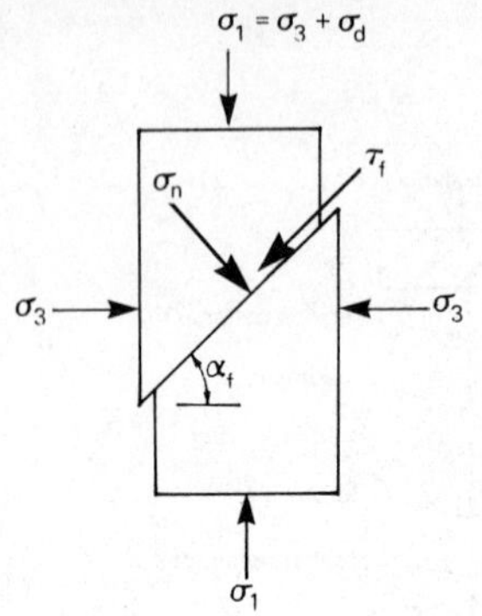

Fig 13.12 Stresses at failure in the triaxial test

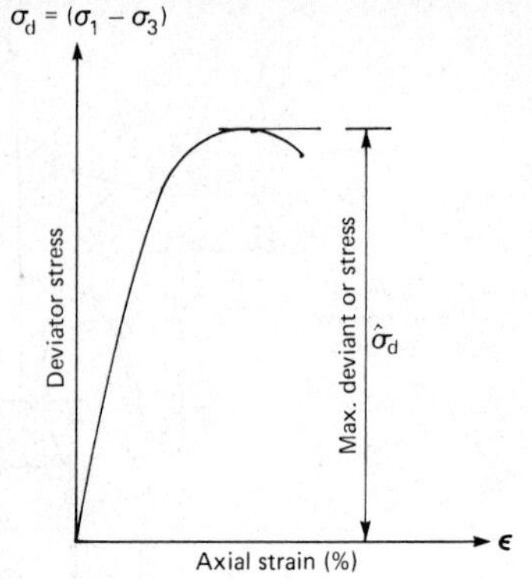

Fig 13.13

Major principal stress, σ_1 = cell pressure + axial deviator stress.
$= \sigma_3 + \Delta\sigma_1$

Axial deviator stress, $\Delta\sigma_1 = \dfrac{\text{axial load}}{\text{cross sectional area } (A)}$

Since the specimen expands laterally as it is compressed axially, the cross-sectional area increases with the axial strain – a standard relationship is usually assumed for this:

$$A = \frac{A_o}{1 - \epsilon}$$

where A_o = initial cross section area

ϵ = axial strain = $\dfrac{\text{change in length}}{\text{original length}}$

The **effective** principal stresses will be

$\sigma'_1 = \sigma_1 - u$

$\sigma'_3 = \sigma_3 - u$

where u = pore pressure

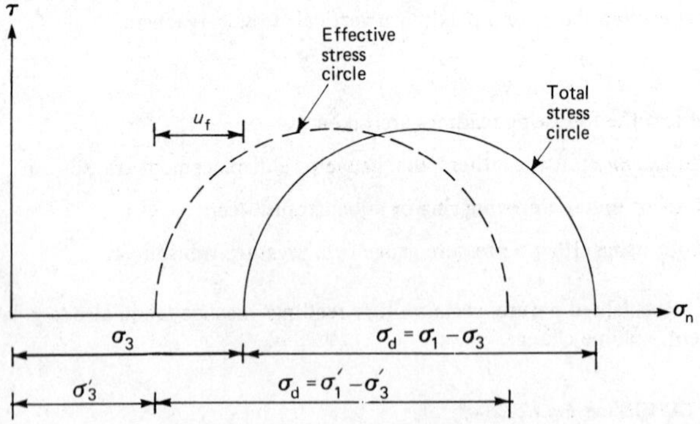

Fig 13.14 Total and effective stress circles

2 For each specimen tested, a plot is drawn of axial stress against axial strain, from which the maximum deviator stress (σ_d) is determined (*Fig 13.13*).

3 For each specimen tested, a Mohr circle is then drawn (*Fig 13.14*) in which

σ_3 = cell pressure

diameter = $\sigma_1 - \sigma_3$ = max. deviator stress (σ_d)

or in terms of effective stresses

σ_3 = cell pressure – pore-pressure at failure

$= \sigma_3 - u_f$

diameter = $\sigma'_1 - \sigma'_3 = \sigma_1 - \sigma_3 = \sigma_d$

4 After at least three Mohr circles have been obtained, a common failure envelope is drawn to obtain the appropriate parameters of cohesion and angle of shearing resistance (*Fig 13.15*).

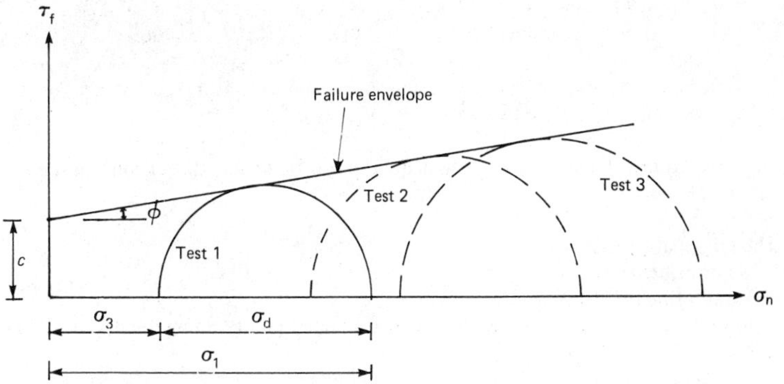

Fig 13.15 Mohr-Coulomb construction to draw failure envelope

TYPES OF TEST

One of the chief advantages of the triaxial test is the capability it offers to control drainage and the rate of strain. The tests summarised below are some of the main ones used that enable procedures to be matched to actual conditions in site problems.

Undrained test. Drainage is prevented throughout the test, so that no dissipation of pore pressure is possible.

Parameters obtained: c_u and ϕ_u

Typical site problem: immediate bearing capacity of foundations in saturated clay.

Consolidated – undrained test. Free drainage is allowed for (usually) 24 hours under cell pressure only to allow the specimen to consolidate or to become saturated. Drainage is then prevented and pore-pressure readings taken during the application of axial load (i.e. shearing stage).

Parameters obtained: c' and ϕ' (i.e. referred to effective stress)
c_{cu} and ϕ_{cu} (i.e. referred to total stress)

Typical site problem: sudden change in load, after an initial stable period, e.g. rapid drawdown of water behind a dam; or where effective stress analysis is required, e.g. slope stability.

Drained test. Free drainage is allowed during a consolidation stage and drainage maintained during the axial loading (which is carried out at a slow rate) so that no increase in pore pressure occurs.

Parameters obtained: c'_d and ϕ'_d

Typical site problem: long-term slope stability

Worked example 13.3 A drained triaxial compression test was carried out on three specimens prepared from the same soil. The results are given below.

Cell pressure (kN/m^2)	100	200	300
Deviator stress at failure (kN/m^2)	224	442	669

Determine the drained values of cohesion and angle of shearing resistance for the soil, assuming zero pore pressure during the test.

In a drained test the pore pressure remains at a constant value throughout; in this case it was zero.

The principal stresses are:

Major principal stress, σ'_1 = cell pressure + deviator stress.

Minor principal stress, σ'_3 = cell pressure.

A Mohr circle is plotted for each specimen's failure point, taking the deviator stress as the diameter (*Fig 13.16*).

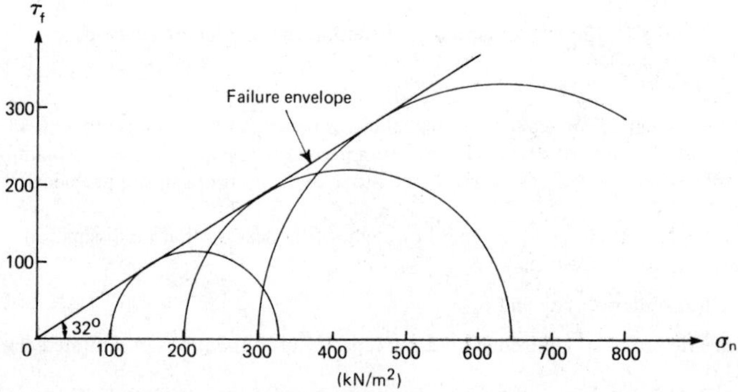

Fig 13.16

The best common tangent drawn to the three circles passes through the origin and slopes at an angle of 32°.

Therefore, the drained shear strength parameters are:

$c'_d = 0$

$\phi'_d = 32°$

Worked example 13.4 The results given below were obtained from a consolidated-undrained triaxial compression test on a clay soil.

Cell pressure (kN/m^2)	100	200	400
Deviator stress at failure (kN/m^3)	248	294	382
Pore pressure at failure (kN/m^2)	13	96	264

Plot the failure envelopes with respect to both total stress and effective stress and determine the values of c_{cu}, $\emptyset_{cu}$, c' and $\emptyset'$.

The Mohr-Coulomb plots are shown in *Fig 13.17*. For the total-stress plot each circle is located by

$\sigma_3 =$ cell pressure;

and $\sigma_1 =$ cell pressure + deviator stress.

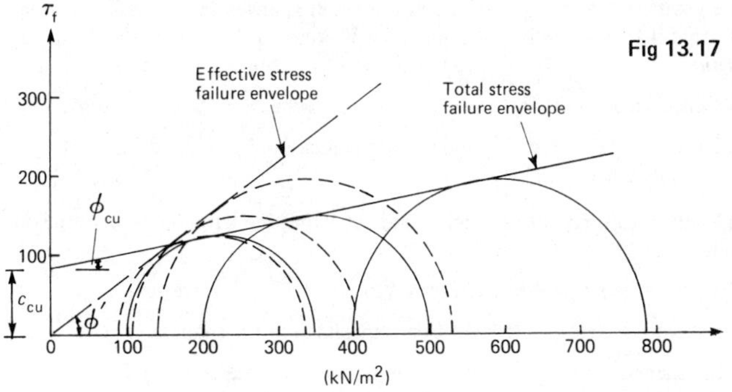

Fig 13.17

The best common tangent to these three circles is the total stress failure envlope, giving the following parameters referred to total stress:

$c_{cu} = $ **84 kN/m²**

$\emptyset_{cu} = $ **11°**

For the effective-stress plot each circle is located by:

$\sigma'_3 = \sigma_3 - u_f =$ cell pressure – pore pressure

$\sigma'_1 = \sigma_1 - u_f =$ cell pressure + deviator stress – pore pressure

Note: $\sigma'_1 - \sigma'_3 = \sigma_1 - \sigma_3$

σ_3	100	200	400
σ_3'	87	104	136
σ'_1	335	398	518

The best common tangent to these three circles is the effective stress envelope, giving the following parameters referred to effective stress:

$c' = 0$
$\phi' = 36°$

TEST EXERCISES

COMPLETION QUESTIONS (answers on page 209)

Complete the following statements by inserting the most appropriate word or words in the spaces indicated.

1 The shear strength of a soil is defined as maximum intensity of . . . . developed on a ruptured surface.

2 The two parameters of shear strength are . . . . and . . . .

3 The shear strength (τ_f) of a soil corresponding to a normal stress of σ_n on the shear plane is given by Coulomb's equation as follows:

τ_f = ... + σ_n . . . .

4 The geometric construction whereby the failure envelope is fitted to a series of principal stress circles (or semi-circles) is termed the . . . . failure theory.

5 In soils such as sands and gravels, the value of the . . . . is zero.

6 In undrained saturated silts and clays, the measured value of . . . . is normally zero.

7 The subscript used with c_u and ϕ_u indicates that the parameters were measured under . . . . conditions.

8 The parameters c' and ϕ' are referred to . . . . stresses.

9 The shear box test is carried out on a specimen in the shape of a . . . .

10 In the shear box test, the load hanger assembly is used to apply the . . . . stress and the screw jack is used to apply the . . . . stress.

11 The triaxial compression test is carried out on specimens in the shape of . . . .

12 In the triaxial test, the cell pressure provides the lateral or . . . . stress and the screw jack increases the vertical or . . . . stress.

13 The difference between the vertical and lateral stresses is called the . . . . stress and is plotted as the . . . . of the Mohr circle.

14 Three main types of triaxial test may be conducted depending on the drainage conditions allowed for the specimens, these are: . . . . , ..., ...

15 Indicate which of the three types of triaxial test would be suitable when measuring parameters in connection with the following problems:

(i) long term slope stability: . . . .
(ii) bearing capacity of a saturated clay: . . . .
(iii) stability of an earth dam after a rapid reduction in the stored water level: . . . .

ESSAY QUESTIONS

1 Explain the meaning of **internal friction** and **cohesion** in terms of the composition of soil. State the **measured** values of these used to define the shear strength of a soil.

2 Briefly describe the Mohr/Coulomb failure theory and its application to the measurement of shear strength.

3 Explain how the drainage conditions associated with a soil mass will influence its shear strength. How is this problem dealt with from the practical point of view when measuring shear strength?

4 Compare the relative advantages and disadvantages of the shear box and triaxial compression tests.

CALCULATION EXERCISES (answers on page 210)

1 The following results were recorded during a shear box test on a silty sand.

Normal load (N)	198	374	551	727	904
Shear load at failure (N)	173	281	356	454	551

The specimen size was 60 mm × 60 mm.
Plot the failure envelope and determine from it the apparent cohesion and angle of shearing resistance.

2 The following results were recorded during shear box tests on a clean dry sand, each specimen size being 250 mm × 250 mm.

Normal load (N)		267	657	1051
Shear load at failure (N)	Peak	246	618	982
	Ultimate	160	400	637

Determine the angles of shearing resistance of the sand corresponding (a) to a dense state and (b) to a loose state.

3 Two specimens of the same sand were tested in a 60 mm shear box, the first being compacted to a dense state and the second prepared in a loose condition. The normal stress was maintained at 150 kN/m^2 in both cases. The data below was recorded during the tests.

TEST 1. DENSE SAND

Horizontal displacement (10^{-2} mm)	0	50	100	150	200	250	350	450	550
*Vertical displacement** (10^{-2} mm)	0	−6	−6	−1	10	22	37	41	42
Shear stress (kN/m^2)	0	67	112	134	142	138	108	99	97

* +*ve* = up (expansion)

TEST B. LOOSE SAND

Horizontal displacement (10^{-2} mm)	0	50	100	150	200	250	350	450	550
*Vertical displacement** (10^{-2} mm)	0	−8	−13	−16	−18	−20	−21	−21	−21
Shear stress (kN/m^2)	0	38	65	82	88	93	94	95	95

* −*ve* = down (compression)

(a) Plot the graphs of shear stress against horizontal displacement and vertical displacement against horizontal displacement for both tests.
(b) Determine the peak and ultimate values of the angle of shearing resistance.
(c) Calculate the estimated percentage dilation or compression in each case, assuming an initial specimen thickness of 20 mm.

4 The results of an undrained shear box test on a clay soil are given below.

Normal stress (kN/m^2)	54	153	251
Shear stress at failure (kN/m^2)	84	110	134

(a) Plot the failure envelope and determine the undrained shear strength parameters c_u and ϕ_u.
(b) Determine the value of the undrained shear strength (c_u) of the soil that would b indicated by an unconfined compression test (i.e. when $\sigma_3 = 0$).
(c) Determine the deviator stress at which a specimen of this soil would fail in a triaxial compression test at a cell pressure 150 kN/m^2.

5 (a) Write down the Coulomb equation for the shear strength of a soil with respect tc effective stress.
(b) The normal stress acting on a given plane in a soil mass is 224 kN/m^2, at the sam point the pore pressure is 75 kN/m^2. If $c' = 15\ kN/m^2$ and $\phi' = 32°$, calculate th shear strength of the soil developed at this point.

6 The following data was recorded during an undrained triaxial compression test on a clay soil.

Cell pressure (kN/m^2)	100	250	400
Deviator stress at failure (kN/m^2)	262	341	420

Using the Mohr–Coulomb theory, draw the failure envelope and determine the apparent cohesion and angle of shearing resistance.

7 The following data was recorded during a consolidated-undrained triaxial compressi test of a clay soil.

Cell pressure (kN/m^2)	150	300	450
Deviator stress at failure (kN/m^2)	250	316	383
Pore pressure at failure (kN/m^2)	52	176	299

Construct the failure envelopes with respect to total and effective stress, and hence determine the parameters c_{cu}, ϕ_{cu}, c' and ϕ'.

8 The following data was recorded during a consolidated-undrained triaxial compression test on a clay soil.

Cell pressure, (kN/m^2)	100	250	400
Deviator stress at failure (kN/m^2)	320	425	530
Pore pressure at failure (kN/m^2)	−33.5	53	140

Construct the failure envelopes with respect to total and effective stress and hence determine the parameters c_{cu}, ϕ_{cu}, c' and ϕ'.

Answers to exercises

Chapter 1 (see page 6)

1 4600;
2 Core, mantle, crust;
3 Core, mantle, mantle, crust.
4 Eras, periods, epochs, ages;
5 Pre-Cambrian, Cambrian, Ordovician, Silurian, Devonian, Carboniferous, Permian, Triassic, Jurassic, Cretaceous, Tertiary, Quarternary;
6 Stratigraphic;
7 System, series;
8 Triassic, Carboniferous, Cretaceous;
9 Principle, superposition;
10 Three from: surface marks, surface borings, cross-bedding (or current bedding), fossil growth position and grading changes;
11 N.W. Scotland, Anglesey;
12 Tertiary, East Anglia and the Londo and Hampshire basins.

Chapter 2 (see page 21)

1 Changing;
2 James Hutton;
3 Wind, rivers, sea, ice, mankind;
4 Mineral composition, physical properties, climate, exposure;
5 9%;
6 Flaky, scree, talus;
7 Solifluction;
8 Exfoliation;
9 Abrasion, ablation;
10 Attrition;
11 Solution, oxidation, hydration, carbonation;
12 $(H)^{+}$, $(OH)^{-}$;
13 Solution;
14 Solution, cementing;
15 Haematite, limonite, goethite;
16 Hydration, orthoclase, pyroxene;
17 Kaolinite;
18 Carbonation, $Ca(HCO_3)_2 \leftrightharpoons CaCO_3 + H_2O + CO_2$;
19 One of: pot-holes, sink holes, swallets, caverns;
20 Calcium carbonate, tufa;
21 Residual;
22 Aeolian, glacial, alluvial, debris, or scree, littoral;
23 Landslips, slides & flows, creep and bulg
24 Youthful, mature, senile;
25 Solution, suspension, saltation;
26 Flood plain or alluvial flat;
27 Meanders, ox-bow lakes or mortlakes;
28 Four from: U-shaped, hanging valley, co arête, horn, roche moutonnée crag and t moraine deposits;
29 Moraine, boulder clay, drumlines, eskers. kames;
30 Wave-cut platform;
31 Longshore drift or littoral drift, spit, bar tombolo.

Chapter 3 (see page 42)

1 Silicon, oxygen;
2 Metallic;
3 Non-metallic;
4 Carbon and silica;
5 Silicates;
6 Lustre;
7 Cleavage;
8 Hardness;
9 Talc, diamond;
10 7;
11 Biotite, muscovite;
12 Perfect, three;
13 Iron, lead;
14 Calcite;
15 Quartz;
16 Acid, coarse;
17 Glassy, fine, basic;
18 Quartz, orthoclase, mica (biotite or muscovite);
19 Clay;
20 Olivine;
21 Amphibole;
22 Evaporites;
23 Porphyrite;
24 Pyroclastic;
25 Consolidation, cementation;
26 Fragments;
27 Solutions;
28 Clastic;
29 Precipitated;
30 Limestone;
31 Thermal, dynamic and regional;
32 Aureole;
33 Quartzite, marble;
34 Dynamic;
35 Slaty, slates;
36 Schistosity or schistose texture;
37 Gneisses.

Chapter 4 (see page 64)

1 Depositional, tectonic, erosional;
2 Beds, bedding planes;
3 Joints;
4 Bed;
5 Strike;
6 Strike;
7 Angle, horizontal;
8 Amount, direction;
9 ↗ + ⊥ ;
10 Scarp (or escarpment), dip slope, cuesta;
11 Mesa;
12 Outlier;
13 Conformable;
14 Unconformity, overstep, overlap;
15 Competent, incompetent;
16 Anticlines, synclines;
17 Axial plane;
18 Overfold, recumbent fold, isoclinal folding;
19 Away from;
20 Fracture, shear;
21 Dip, hade, throw
22 Hades;
23 Reversed;
24 Two from: wrench, tear, transform, transcurrent;
25 Strike;
26 50 000;
27 Solid, drift;
28 10.

Chapter 5 (see page 77)

1 Hydrology;
2 Hydrological;
3 Precipitation, evaporate;
4 Infiltration;
5 Run-off;
6 Any three from: ground slope, vegetation, temperature, rainfall, porosity, permeability, discontinuities, stream occurrence;
7 Vadose;
8 Phreatic (or gravitational), phreatic;
9 Atmospheric;
10 Perched water table;
11 Suction (or capillary suction);
12 Impermeable or impervious, permeable, pervious;
13 Will, will not;
14 Catchment, catchment, watershed;
15 Isotropic, anisotropic;
16 Groundwater;
17 Bourne;
18 Artesian, piezometric;
19 Artesian, artesian;
20 Artesian, confined aquifer;
21 Freshwater/saline water;
21 1.025;
22 N.W. Scotland, Wales, the Lake District, S.W. England;
23 Three from the list given on page 00.

Chapter 7 (see page 94)

1 BS 5930: 1981;
2 These are given on page 00;
3 Geological, topographical, information, economic;
4 Four from: trial pits, headings, hand auger, shell and bailer, rotary auger, core drilling;
5 Undisturbed;
6 Piston sampler;
7 Bulk density;
8 Standard penetration;
9 Dutch cone;
10 Shear strength;
11 Shear strength, compressibilit

Chapter 8

Completion questions (see page 114)

1 *Three* from: particle size and shape, grading, plasticity indices, mineral composition and colour;
2 *Three* from: bedding, discontinuities, moisture content, compactness, weathered condition;
3 425;
4 Particle size, plasticity;
5 CLAY of extremely high plasticity, well-graded silty GRAVEL, very clayey SAND (clay of low plasticity);
6 *Five* from: particle size, grading, amount of fines, plasticity, dilatancy, roughness, dry strength, penetration resistance, soil structure;
7 Diameter;
8 10%;
9 Maximum size of the smallest 60% (D_{60});
10 Less than 3.0;
11 To flow as a liquid;
12 Liquid, plastic;
13 High;
14 Silts, clays.

Calculation questions (see page 115)

1 Very silty uniform fine SAND.
2 0.052, 0.165, 0.38; Gap-graded gravelly silty SAND.
3 49%, CLAY of intermediate plasticity.
4 (a) CH, (b) MV, (c) MIO, (d) SCH, (e) GPM.

Chapter 9

Completion questions (see page 130)

1 Solid, liquid and gas;
2 Flakiness;
3 Negative;
4 Suction;
5 Equilibrium;
6 Plasticity, cohesion;
7 Volume of solids;
8 Mass of solids;
9 Volume of voids;
10 Mass of solids;
11 Total mass;
12 $(G_s + S_r e)\rho_w/(1 + e)$;
13 ρ_w;
14 $1 + m$;
15 2.64, 2.72;
16 1.5, 2.2;
17 Compaction;
18 Maximum dry density, optimum moisture content;
19 Air-void;
20 Air-void.

Calculation questions (see page 131)

1 0.639, 0.755, 0.096;
2 (a) 2.039 Mg/m^3, 1.808 Mg/m^3, (b) 0.493, 0.330, (c) 0.701, 0.099;
3 (a) 2.074 Mg/m^3; (b) 1.890 Mg/m^3;
4 0.583;
5 0.563; 16.76 kN/m^3, 20.29 kN/m^3;
6 (a) 0.667, (b) 0.237, 0.952; (c) 0.019, (d) 15.77 Mg/m^3, 19.70 Mg/m^3;
7 0.820;
8 (a) 1.917, 12.8%, (b) 4%.

Chapter 10

Completion questions (see page 146)

1 Velocity;
2 *Aki*;
3 1×10^{-3} to 1×10^{-4};
4 1×10^{-7};
5 Sands and gravels;
6 Silts and clays;
7 $\log_e(h_1/h_2)$, $A/t_2 - t_1$);
8 Quicksand;
9 $(G_s - 1)$, $(1 + e)$;
10 Flow lines;
11 Equipotential lines;
12 Flow, equipotential;
13 At right angles;
14 N_f/N_e.

Calculation questions (see page 148)

1 1.4×10^{-3} m/s;
2 5 min 40 s, 3.4s;
3 8.1×10^{-7} m/s;
4 4.1×10^{-4} m/s;
5 5.3×10^{-4} m/s;
6 1.07 m;
7 0.071 m^3/h;
8 (a) 0.35 m^3/h, (b) 15 kN/m^2, 13 kN/m^2, 8 kN/m^2.

Chapter 11

Completion questions (see page 167)

1 Pore fluid;
2 Soil fabric (or intergranular contacts);
3 $\sigma = \sigma' + u$;
4 $\gamma' z$ (γ' = submerged density);
5 K_a, K_p, K_o;
6 $\nu/(1 - \nu)$, $1 - \sin \phi'$;
7 $K_a = (1 - \sin \phi')/(1 + \sin \phi')$ or $K_a = \tan^2(45° - \phi'/2)$, $K_p = (1 + \sin \phi')/(1 - \sin \phi')$ or $K_p = \tan^2(45° + \phi'/2)$;
8 $p_a = K_a \sigma'_z - 2c'\sqrt{K_a}$, $p_p = K_p \sigma'_z + 2c'\sqrt{K_{pi}}$;
9 Wedge theory;
10 Gravity, seepage.

Calculation questions (see page 168)

1. At $Z = 2.0$ m, $\sigma' = 38.0$ kN/m², $\sigma =$ 38.0 kN/m²; at $z = 5.0$ m, $\sigma' = 69.2$ kN/m², $\sigma = 98.6$ kN/m²; at $Z = 9.5$ m, $\sigma' = 106.1$ kN/m², $\sigma = 179.7$ kN/m²;
2. (a) 79.2 kN/m², 98.8 kN/m², (b) 54.25 kN/m², 103.3 kN/m²;
3. (a) 64.9 kN/m², (b) 31.8 kN/m²;
4. (a) 0.577, 0.406, 2.46, (b) 0.792, 0.656, 1.52;
5. (a) 278 kN/m, (b) 554 kN/m
6. 1244 kN/m acting at 4.75 m above the base.
7. 408 kN/m acting at 3.48 m above the base;
8. (a) 451 kN/m acting at 2.57 m above the base, (b) 480 kN/m acting at 3.75 m above the base;
9. 408 kN/m;
10. 770 kN/m; 773 kN/m, 694 kN/m.

Chapter 12

Completion questions (see page 183)

1. *Five* from: compaction, consolidation, elastic compression, moisture movement, vegetation transpiration, groundwater lowering, temperature changes, piping and scouring, subsidence;
2. Elastic;
3. Compaction;
4. Consolidation;
5. H_o, $(1 + e_o)$;
6. $m_v \Delta\sigma' H_o$;
7. Oedometer;
8. Normally-consolidated;
9. Compression;
10. $T_v d^2/c_v$;
11. Over consolidated;
12. Over consolidated;
13. Expand or dilate.

Calculation questions (see page 185)

1. 10 mm;
2. 0.092, 103 mm;
3. 338 mm;
4. (a) 59 mm, (b) 1.28 years
5. 291 mm, 53.8 years.

Chapter 13

Completion questions (see page 200)

1. Shear stress;
2. Angle of shearing resistance, apparent cohesion;
3. c, tan $\emptyset$;
4. Mohr-Coulomb;
5. Apparent cohesion (c);
6. Angle of shearing resistance ($\emptyset_u$);
7. Undrained;
8. Effective;
9. Rectangular prism;
10. Normal, shear;
11. Cylinders;
12. Minor principal, major principal;
13. Deviator stress, diameter;
14. Undrained, consolidated-undrained, drained;
15. Drained, undrained, consolidated-undrained.

Calculation questions (see page 201)

1. 26 kN/m², 28°;
2. (a) $\phi'_{dense} = 43°$, (b) $\phi'_{loose} = 31°$;
3. (b) $\phi'_{peak\,2} = 40°$, $\phi'_{ultimate} = 32°$; (c) 2.1%; 1.05%.
4. (a) 72 kN/m², 14°, (b) 92 kN/m², (c) 280 kN/m²
5. 164 kN/m²;
6. 85 kN/m², 12°;
7. 76 kN/m², 10½°, 0, 34°;
8. 96 kN/m², 15°, 30 kN/m², 27°.

Index